Digital Signal Processing
A Laboratory Approach Using PC-DSP

OKTAY ALKIN
*Southern Illinois University
at Edwardsville*

Prentice-Hall International, Inc.

This edition may be sold only in those countries to which
it is consigned by Prentice-Hall International. It is not to
be re-exported and it is not for sale in the U.S.A., Mexico,
or Canada.

The author and publisher of this book have used their best efforts in preparing this
book. These efforts include the development, research, and testing of the theories and
programs to determine their effectiveness. The author and publisher make no warranty
of any kind, expressed or implied, with regard to these programs or the documentation
contained in this book. The author and publisher shall not be liable in any event for
incidental or consequential damages in connection with, or arising out of, the
furnishing, performance, or use of these programs.

© 1994 by Prentice-Hall, Inc.
A Paramount Communications Company
Englewood Cliffs, New Jersey 07632

All rights reserved. No part of this book may be
reproduced, in any form or by any means,
without permission in writing from the publisher.

Printed in the United States of America

10 9 8 7 6 5 4 3 2 1

ISBN 0-13-079542-9

Prentice-Hall International (UK) Limited, *London*
Prentice-Hall of Australia Pty. Limited, *Sydney*
Prentice-Hall Canada Inc., *Toronto*
Prentice-Hall Hispanoamericana, S.A., *Mexico*
Prentice-Hall of India Private Limited, *New Delhi*
Prentice-Hall of Japan, Inc., *Tokyo*
Simon & Schuster Asia Pte. Ltd., *Singapore*
Editora Prentice-Hall do Brasil, Ltda., *Rio de Janeiro*
Prentice-Hall, Inc., *Englewood Cliffs, New Jersey*

To my parents

Contents

Preface viii

1 Introduction **1**

    1.1 What Is PC-DSP? 2

    1.2 A Guided Tour 3

2 Discrete-time Signals **10**

    2.1 Elementary Discrete-time Signals 13

    2.2 Classification Methods for Discrete-time Signals 16

    2.3 Elementary Operations on Discrete-time Signals 20

    2.4 Symmetry Properties of Discrete-time Signals 26

    2.5 Discrete-time Sinusoids Revisited 28

3 Discrete-time Systems **31**

    3.1 Linear and Time-invariant Systems 32

    3.2 Signal-system Interaction 35

    3.3 System Function Concept 43

    3.4 Difference Equations 46

    3.5 Stability and Causality 53

4 Frequency-domain Analysis 57

4.1 Discrete-time Fourier Transform 58
4.2 Properties of the Discrete-time Fourier Transform 60
4.3 Symmetry Properties of the DTFT 67
4.4 Frequency-domain Analysis of Systems 69
4.5 Linear-phase Systems 70
4.6 Discrete Fourier Transform 72
4.7 Properties of the Discrete Fourier Transform 76

5 Sampling and Reconstruction 82

5.1 Sampling of Continuous-time Signals 84
5.2 Reconstruction 87
5.3 Quantization 90
5.4 Changing the Sampling Rate 95

6 The z-Transform and Discrete-time Structures 101

6.1 Basic Concepts 101
6.2 Inverse z-Transform 108
6.3 Discrete-time Structures 111

7 IIR Filter Design 113

7.1 Overview of IIR Filter Design Methods 114
7.2 IIR Filter Specifications 115
7.3 IIR Filter Design Using Analog Prototypes 118
7.4 Analog to Discrete-time Conversion 118
7.5 Obtaining Analog Prototype Specifications 128
7.6 Designing IIR Filters Using PC-DSP 129

8 FIR Filter Design 132

8.1 Fourier Series Design of FIR Filters 133
8.2 Frequency Sampling Design 144
8.3 Parks–McClellan Algorithm 149

Contents

Appendix A: PC-DSP Reference 151

A.1　Basic Concepts　152

A.2　Function Reference　156

A.3　Data File Formats　189

Appendix B: Analog Filter Design 190

B.1　Butterworth Filters　191

B.2　Chebyshev Filters　196

B.3　Inverse Chebyshev Filters　202

B.4　Elliptic Filters　205

B.5　Frequency Transformations　206

References 209

Index 211

Preface

Over the past decade, teaching methods used in courses that involve linear systems and signal processing have evolved significantly beyond the "lecture-only" format. Experience has shown that, for effective teaching of signal processing, computer use in the form of laboratory exercises and computer-assisted homework must be made an integral part of the course. Students learn difficult concepts most effectively when those concepts can be presented visually, in a form suitable for experimentation. Slightly modifying the infamous cliché, a graph on the computer screen is sometimes worth a thousand words in the textbook.

This textbook and its accompanying software are intended for use as supplements in junior- and senior-level undergraduate courses that involve linear systems and signal processing. They can also be used for self-study in conjunction with one of the conventional DSP texts available.

The textbook is divided into eight chapters that parallel the coverage of most undergraduate-level DSP books. It contains a large number of exercises that can be performed using the software included. These exercises can be used in a formal laboratory setting or can be assigned as part of the homework for the course. Each exercise attempts to illustrate and reinforce a particular concept. Coverage of the related DSP theory is provided leading up to each exercise. Theoretical derivations, when included, are not intended to be mathematically rigorous or to replace the derivations in the main textbook of the course. Their main purpose is to ensure that the requirements of the exercises that follow and the motivation for performing them are clear and that the notation is understood. Since traditional DSP texts differ slightly from each other in the notation used, to not have theoretical coverage in this book would have made some exercises difficult to understand when they were used in conjunction with some textbooks.

A student edition of PC-DSP version 2 is included with this textbook. PC-DSP was written with the intent to provide the student with an integrated signal-processing environment in which realistic digital signal-processing problems can be solved. It employs an interactive user interface designed to minimize the time needed for learning the operation of the program. No programming knowledge is required on the part of the user, although basic familiarity with personal computers and the disk operating system (DOS) is helpful. One can generate, analyze, and process signals, compute transforms, and design and analyze filters by simply going through a set of menu choices, without the

Preface

need to remember any command syntax. A context-sensitive on-line help feature is also provided for added ease of use.

During the several years that the early version of PC-DSP has been in use, a good deal of feedback was generated by instructors and students using the program. As a result, version 2 contains significant enhancements. The user interface is more intuitive and easier to learn. A number of new functions have been added, and existing functions have been enhanced, particularly in the areas of waveform synthesis, data-file conversions, and filter design. The macro language has been expanded to make it more versatile.

The software will work on a personal computer operating under MSDOS operating system version 3.0 or higher, with at least one disk drive, 640 K-bytes of random-access memory and graphics display capability. It will take advantage of a floating-point math coprocessor (if one is installed) to speed up floating-point math operations. Although it is possible to use the program on a floppy-disk-based system, a fixed-disk is strongly recommended.

This book consists of eight chapters and two appendixes. Chapter 1 provides an introduction. The basic philosophy and the format of the text and the software are explained, and a brief guided tour of the software is provided. In Chapter 2, discrete-time signal concepts are explored. Symmetry properties of signals and periodicity of discrete-time sinusoids are also covered in this chapter. The discrete-time system concept is introduced in Chapter 3. Exercises are provided to elaborate on three forms of description for discrete-time linear and time-invariant systems using convolution, system function, and difference equations. Chapter 4 concentrates on frequency-domain transforms. The discrete-time Fourier transform (DTFT) is introduced, and its important properties are explained. The DTFT-based system function concept is further detailed. The discrete Fourier transform (DFT) is also introduced in this chapter. A number of exercises are provided to illustrate the ties between the DTFT and the DFT. The issue of sampling analog signals, and altering the sampling rate on discrete-time signals is covered in Chapter 5. Reconstruction of signals from their sampled versions is also discussed in this chapter. In addition, a brief treatment of uniform and nonuniform quantization is provided. In Chapter 6, the z-transform is covered. Its relationship to the DTFT is discussed. This chapter also contains a brief coverage of real-time implementation structures based on z-domain system functions. Design methods for infinite-impulse-response (IIR) filters are discussed in Chapter 7. The steps involved in the design of IIR filters based on analog prototypes are detailed, and a large number of exercises are provided to aid in understanding various methods. Computer-aided analysis and simulation methods for IIR filters designed using PC-DSP, as well as the student's pencil-and-paper designs, are also explained. Chapter 8 treats the design and analysis of finite-impulse-response (FIR) filters. Fundamentals of the three design techniques employed are briefly explained. Exercises are provided to help in understanding Fourier-series design of FIR filters, the Gibbs phenomenon, window functions, and the frequency-sampling design method. Appendix A provides a reference guide for the software. Descriptions of PC-DSP functions are given. Formats of binary data files used by PC-DSP are provided so that the same data files can be read by user programs if desired. Most of this information is also contained in the on-line help system of the software. In Appendix B, the design of Butterworth, Chebyshev, and elliptic filters is discussed. This is included for completeness of the filter design coverage and to provide insight into the filter design methods used by PC-DSP for the interested reader. Although no analog design exercises are given, analog filters can be designed using PC-DSP.

I would like to express my appreciation to those who contributed to the successful completion of this project. The following reviewers have provided helpful comments and suggestions: Professor Bernard Hutchins (Cornell University); Professor William Tranter (University of Missouri-Rolla); and Professor VanLandingham (Virginia Tech). Thanks are also due to Professor Ken Fyfe of University of Alberta and to Dr. Gerald Burnham of Texas Instruments Inc., who have used the earlier edition in their courses and provided feedback on ways it could be improved. I am grateful to my students in EE 436 who have worked the exercises in the text, tested development versions of the software, and tried to "catch the bugs." Finally, I would like to thank my wife, Esin, for her patience and understanding throughout this project.

1

Introduction

*If I hear it, I will forget it.
If I see it, I will remember it.
If I do it, I will understand it.*
—Chinese proverb

This textbook and software package is intended to be used as a supplement in junior- and senior-level courses that cover linear systems theory and signal processing. The software provided with this text is the student edition of PC-DSP version 2. It is designed to take the tedium out of performing common signal-processing operations and exploring fundamental concepts encountered in this subject area. This chapter will serve as an introduction to the basic philosophy of the text and software. A brief guided tour of the software will also be given.

The requirements for using the included software are as follows:

- An IBM PC or 100% compatible computer operating under the Microsoft Disk Operating System (MS-DOS) version 3.0 or higher.
- At least 512 K-bytes of RAM (random-access memory) installed in the system. 640 K-bytes is recommended if user-written programs are to be interfaced to PC-DSP for advanced applications.
- Graphics display capability (CGA, EGA, VGA graphics adapter).

- A fixed disk drive is strongly recommended. Although it is possible to configure the program to run on a floppy-disk-based system (at least 1-MB total storage capacity required), this results in significantly slower operation due to the use of overlays and data files.
- A math coprocessor is optional. If it is present in the system, PC-DSP will automatically detect it and use it to speed up floating-point calculations.
- The use of a mouse, although not required, greatly simplifies navigating through the menu structure of the program.
- A printer is not required, but is recommended for obtaining hard copies of tabulations and graphics. To obtain printer graphics, you will also need the proper graphics driver program for your printer (GRAPHICS.COM or similar).
- For proper operation of PC-DSP, it is recommended that the following two lines be placed in the CONFIG.SYS file of the disk operating system:

```
Files = 20
Buffers = 20
```

To install the software on a computer, refer to the instructions that are included on the distribution disk(s) in the file named 'README.TXT'.

1.1 WHAT IS PC-DSP?

PC-DSP (personal computer digital signal processing) is an interactive, menu-driven software package used for the analysis, design, and implementation of discrete-time signals and systems. It can be considered a special-purpose calculator geared toward performing digital signal-processing operations such as waveform synthesis and manipulation, Fourier transforms, convolution, correlation, filter design, analysis and implementation, power spectrum estimation, and graphics.

PC-DSP program code is contained in a number of files with the DOS file name extension 'EXE'. The file PCDSP.EXE is the shell that implements the components used for the display and for interaction with the user. These components include the menu structure, the status bar, dialog boxes for data entry and for tabulation of results, a progress monitor to provide information about the computations performed, an editor for developing macros, and on-screen graphics. Other files with the 'EXE' extension are overlays. They contain code for the numerical algorithms.

In recent years, parallel to the developments in digital signal processing, the increased availability and affordability of personal computers and workstations have led educators into exploring new and better ways of integrating computer use into the curriculum. In most engineering disciplines, the breadth of the material to be covered has increased significantly in the last 10 or 15 years, yet the duration of a typical undergraduate degree program remains unchanged. As a result, educators are faced with the problem of giving their students a sufficient background in a large number of areas in limited time. The effective use of educational software tools seems to be one possible solution to this dilemma.

Sec. 1.2   A Guided Tour   3

Today, a large number of DSP-related computer programs are available commercially. Although most of these commercial programs are adequate for researchers and practicing engineers, they do not necessarily provide the best means for learning the theory of the operations performed. For example, a filter design package that accepts specifications for the desired filter behavior and produces the coefficients of the optimum filter provides very little insight into the design procedure. A more instructive method would be to divide the problem into its basic steps and provide functions to perform each step. As a specific example, an infinite-impulse-response (IIR) bandpass filter could be designed with the following steps in PC-DSP:

- Convert the specifications of the desired IIR bandpass filter to the specifications of an appropriate analog bandpass prototype.
- Convert the specifications of the analog bandpass prototype to the specifications of an analog low-pass prototype. Manually design the analog low-pass prototype.
- Design the analog low-pass prototype filter using the computer.
- Analyze the analog low-pass prototype. Is it what was expected?
- Determine the parameter values to be used in the frequency transformation to convert the analog low-pass prototype to an analog bandpass prototype.
- Apply the necessary frequency transformation using the computer. Afterward, analyze the resulting analog bandpass prototype filter.
- Determine parameter values to be used in converting the analog bandpass filter to a discrete-time bandpass filter.
- Convert the analog prototype to a discrete-time filter using the desired conversion technique. Analyze the resulting discrete-time filter.

If, on the other hand, quick design of a filter is needed for exploring a DSP concept other than filter design, a *do-it-all* type of IIR filter design function is also available. Similar examples can be developed for other subjects covered in a typical DSP course.

1.2 A GUIDED TOUR

This section will provide a basic introduction to PC-DSP by illustrating the use of a few of its functions. Specifically, we will perform the following steps using the program:

1. Generate a sinusoidal discrete-time signal. Graph and tabulate this signal on the screen.
2. Obtain a new signal by squaring each sample of the signal generated in step 1. Tabulate and graph this new signal.
3. Compute the discrete-time fourier transform (DTFT) of the signal obtained in step 2. Tabulate and graph the transform.
4. Quantize the signal obtained in step 2 to eight quantization levels. Tabulate and graph the resulting signal.

5. Filter the quantized signal using a three-tap moving-average filter. Tabulate and graph the output signal of the filter.

Note that the purpose of this section is to familiarize the reader with the operation of the program rather than of the specific signal-processing operations performed in the preceding steps. For detailed information on specific functions of PC-DSP, consult Appendix A or the on-line help guide.

Run the program by typing *PCDSP* at the operating system prompt and pressing the *enter* or *return* key. Upon entry into the program, press the space bar to remove the copyright and version information window.

The PC-DSP main menu screen consists of a number of components. The top row of the screen is occupied by a menu bar that lists the main sections of PC-DSP. The status line is the bottom row of the screen. It contains information about special keys for obtaining on-line help, graphing or tabulating the last sequence computed, gaining access to the menu structure, and exiting the program. Also, the available memory is displayed at the lower right corner of the screen. Right above the status bar, there is a progress monitor. This is a TTY-type scrolling window that displays information about the computations performed.

Step 1. If a mouse-type pointing device is available, position the mouse pointer on the menu item *Sequences*, and press the left mouse button. A pull-down menu should appear. Select *Generate sequence* and *Formula-entry method*. If a mouse is not available, the key sequence *Alt-S, G, F* accomplishes the same thing. (Note that these are the highlighted letters of menu items.) At this point, a data-entry form should appear on the screen with the title *Generate Sequence From A Formula*. (See Fig. 1.1.) This data entry form consists of four fields. Any field can be made current by pressing the left mouse

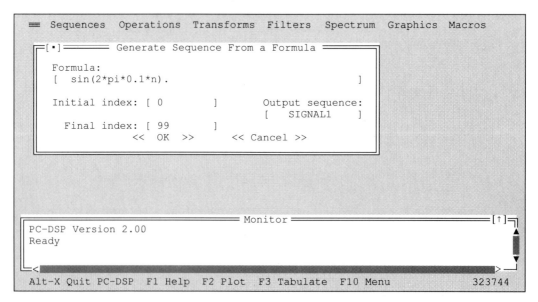

Figure 1.1 Data-entry form for step 1.

Sec. 1.2   A Guided Tour   5

button while the mouse pointer is on that field. Alternatively, the tab key can be used to select a field. To generate a discrete-time sinusoidal signal, make the following entries:

    Formula: `sin(2*pi*0.1*n)`
    Initial index: `0`
    Final index: `99`
    Output sequence: `SIGNAL1`

At this point, press the button labeled *OK* to perform the specified operation. Do this by either pressing the left mouse button while the mouse pointer is on the button labeled *OK* or using the key combination *Alt-O*. This creates a new signal with the name *Signal1*. To tabulate its samples, press the *F3* key, or click on the label *Tabulate* on the status bar. A new window appears with samples of the signal tabulated in terms of their real and imaginary parts. To move up and down in the table, use the directional keys or click on the scroll bars to the right of signal samples. A graph of the generated signal can be obtained by pressing the *F2* key or by clicking on the label *Plot* on the status bar. After viewing the graph, press the space bar to remove it.

Step 2.   From the top-level menu, select *Operations*. Afterward, select *Nonlinear operations* and *Square*. A data-entry window appears with the title *Compute Square*, as shown in Fig. 1.2. Next to the *Input sequence* field is a little button labeled with a down arrow. This indicates that the program expects the name of an existing signal in this field. Either click on this little button or press the down-arrow key while the *Input sequence* field is current. A list of available signals is presented, and the signal generated in step 1 should appear in this list. Select it as the input signal. Afterward, move to the *Output*

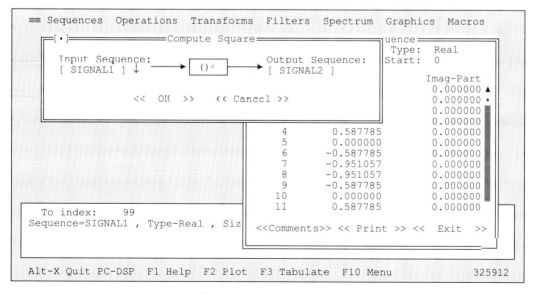

Figure 1.2   Data-entry form for step 2.

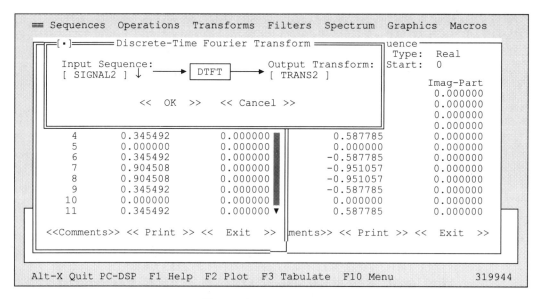

Figure 1.3 Data-entry form for step 3.

sequence field, and type *Signal2*. The fields on the data-entry form should have the following entries:

Input sequence: SIGNAL1
Output sequence: SIGNAL2

Press the button labeled *OK* to perform the specified operation. The resulting signal *Signal2* can be tabulated and graphed as explained in step 1. Position tabulation windows for the two signals side by side and compare corresponding samples. Tabulation windows can be moved by positioning the mouse pointer on the window title and moving the mouse with the left mouse button pressed. Alternatively, the key combination *Ctrl-F5* can be used.

Step 3. From the top-level menu select *Transforms*; then select *DTFT*. This causes the data-entry form with the title *Discrete-Time Fourier Transform* to be displayed, as shown in Fig. 1.3. Make the following entries:

Input sequence: SIGNAL2
Output transform: TRANS2

Press F2 to obtain the magnitude and phase plots shown in Fig. 1.4

Step 4. From the top-level menu, select *Operations*; then select *Nonlinear operations* and *Quantize*. The data-entry form with the title *Quantize Sequence* should be displayed, as shown in Fig. 1.5. Make the following entries:

Sec. 1.2   A Guided Tour

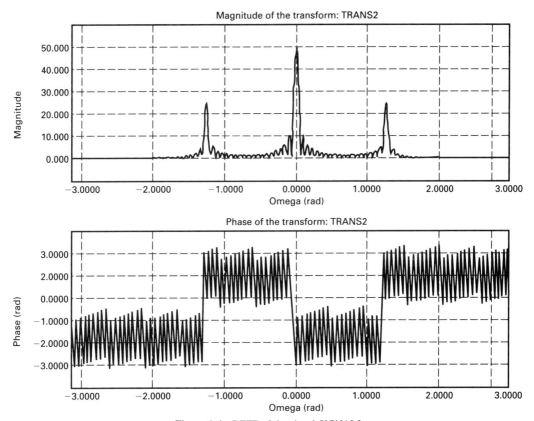

Figure 1.4   DTFT of the signal SIGNAL2.

> Input sequence: SIGNAL2
> Output sequence: SIGNAL3
> Lower limit: 0
> Upper limit: 1
> Number of levels: 8

Tabulate and graph the resulting sequence *Signal3* as in previous steps.

Step 5.   A three-tap moving-average filter is characterized by a difference equation

$$y(n) = \frac{1}{3}[x(n) + x(n-1) + x(n-2)]$$

Select *Operations*; then select *Processing functions* and *Difference equation*. The data-entry form with the title *Difference Equation Solver* should be displayed, as shown in Fig. 1.6. Make the following entries:

8 Introduction Chap. 1

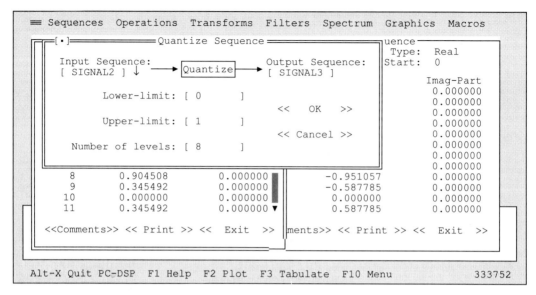

Figure 1.5 Data-entry form for step 4.

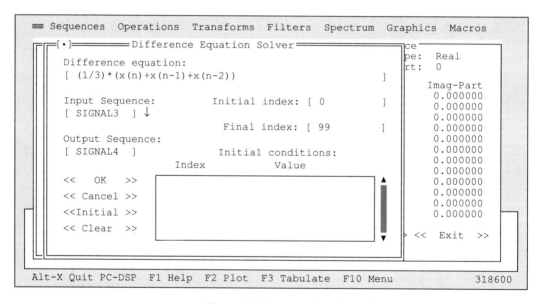

Figure 1.6 Data-entry form for step 5.

Difference equation:	(1/3)* (x(n) + x(n - 1) + x(n - 2))
Input sequence:	SIGNAL3
Output sequence:	SIGNAL4

Sec. 1.2 A Guided Tour

Initial index: 0
Final index: 99

The solution of the difference equation can be graphed with the *F2* key (see Fig. 1.7).

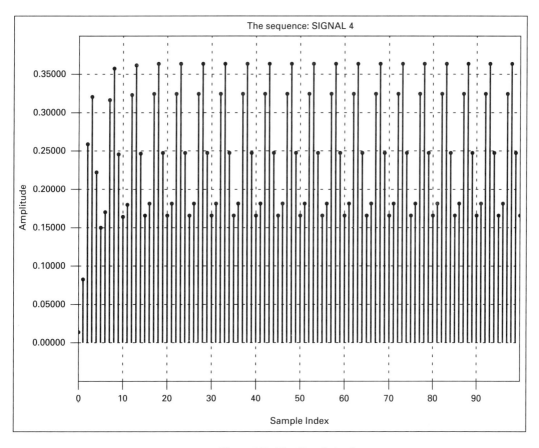

Figure 1.7 The filtered signal.

2

Discrete-time Signals

> The purpose of computing is insight, not numbers.
> —R. W. Hamming, "Numerical Methods for Scientists and Engineers," 1973

In this chapter, we will review the fundamentals of discrete-time signals. A brief introduction to the concepts of discrete-time systems that operate on these discrete-time signals will also be given; a more complete and formal treatment of the systems theory will be deferred until Chapter 3. It would thus make sense to start by defining these concepts and establishing the terminology for use in later parts of the book. Consider the analog system example shown in Fig. 2.1. The system is a physical entity that serves a specific purpose. It might be an electrical circuit that consists of resistors, capacitors, inductors, transistors, integrated circuits, and the like, or perhaps a mechanical system consisting of springs, gears, and other mechanical components. The input to the system is the time variation of some physical quantity, which might be an electrical voltage or a current, the tension on a spring, or the variation of temperature. In systems theory, any of these input signals can be modeled with a mathematical function of time, $x(t)$. The system's task is to somehow transform the input signal into an output signal, which may or may not be of the same type as the input signal. As an example, the system might produce a voltage signal at its output in response to temperature variations at its input. Another system might take the displacement of a lever as input and produce a voltage signal in response. In any

Chap. 2   Discrete-time Signals   11

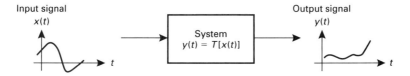

Figure 2.1 Continuous-time system working on continuous-time signals.

case, the output signal of the system can also be modeled with a mathematical function, as we have done with $y(t)$.

In contrast to this analog system example, consider the discrete-time system in Fig. 2.2. In this case, the system is a digital computer that can process numbers. The input signal that is applied to the digital computer is an indexed sequence of numbers in a form suitable for use by the computer and is represented by a mathematical series $x(n)$. (*Note*: The terms *signal* and *sequence* will be used interchangeably throughout the text.) It is possible to draw an analogy from a spreadsheet program and to think of the discrete-time input signal as a set of numbers stored in one column of the spreadsheet. Given the index of a particular cell within that column, the value of the element stored in that cell can be obtained. For example, $x(85)$ refers to the value contained in cell 85 or, in signal-processing terms, the 85th *sample* of the sequence x. (Note that the term *sample* is related to the act of sampling a continuous-time signal to obtain a discrete sequence of numbers. Even though not all discrete-time signals are obtained in this way, we still use this word to refer to each number in a discrete-time signal.) If the index value is outside the range of existing spreadsheet cells, the value of the corresponding sample is assumed to be zero by default; for example, if the spreadsheet is set up to hold 100 samples indexed 0 through 99, $x(105)$ or $x(-20)$ would still be legitimate, and both would be equal to zero. This kind of thinking allows us to treat any discrete-time signal as infinitely long, even though sample values outside a particular range of the index might be equal to zero for the so-called finite-length sequences.

In digital signal-processing literature, it is customary to represent a discrete-time signal graphically as a set of vertical lines placed on a horizontal axis holding the values of the index n. (See Fig. 2.3.) The height of each vertical line represents the value of the corresponding sample of the sequence. Alternative forms of description are also utilized.

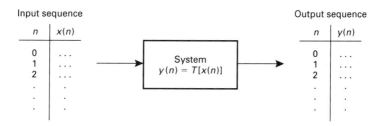

Figure 2.2 Discrete-time system working on discrete-time signals.

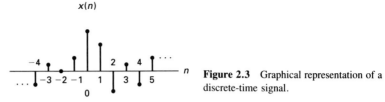

Figure 2.3 Graphical representation of a discrete-time signal.

The signal can be described by means of a mathematical formula. As a particular example, consider the following:

$$x(n) = 0.7n^2 + \sin(0.2\pi n).$$

For any specified integer value of n, the corresponding sample amplitude can be obtained from the formula. Yet another method of describing a discrete-time signal is to tabulate values of its samples. We will adopt a tabular format used in some earlier textbooks.[21] In this format, amplitudes of signal samples are listed between braces, separated from each other by commas. The location of the sample $x(0)$ is indicated by an up arrow (↑) on the next line; that is,

$$x(n) = \{\ldots, x(-2), x(-1), x(0), x(1), x(2), \ldots\}$$
$$\uparrow$$

An ellipsis (...) just after the left brace or just before the right brace indicates the existence of more samples in the respective direction. The absence of an ellipsis on either side indicates that there are no more significant samples in that direction or, equivalently, all samples not shown are zero valued.

The analytical definition of a discrete-time signal is analogous to a programming structure that takes an integer argument and returns a floating-point number based on the value of this integer argument. As an example that uses the C programming language, consider the following piece of code:

```
float MySignal ( int Argument )
{
  if (Argument == 0)
    return 1.7;
  else if (Argument == 1)
    return -0.6;
  else if (Argument == 2)
    return 2.3;
  else
    return 0.0;
}
```

The function *MySignal()* represents an infinite-length sequence that has only three nonzero samples. In tabular form,

$$MySignal(n) = \{1.7, -0.6, 2.3\}$$
$$\uparrow$$

The function can be called with any integer value of the argument, although it returns a value of zero for all except three specific values of it. We will refer to this analogy in later sections of this chapter when considering elementary signal operations. In the case of complex-valued discrete-time signals, complex values can be returned either by using the readily available complex data type of the language (for example, FORTRAN) or by defining a complex data structure (for example, C, Pascal, and some compiled BASICs).

In the continuous-time system example of Fig. 2.1, the actual system would be a physical entity that might consist of electrical, mechanical, electromechanical, or hydraulic components. On the other hand, our mathematical model of the system is a formula or an algorithm that transforms the function $x(t)$ into the function $y(t)$. In the discrete-time system example of Fig. 2.2, the system is a digital computer working with electrical signals that consist of binary pulses, and our model is again a mathematical formula or algorithm that transforms the sequence $x(n)$ into the sequence $y(n)$. Once appropriate mathematical models are found for the system and its input and output signals, the analysis becomes independent of the physical nature of the problem. In other words, the mathematical functions $x()$ and $y()$, and the functional relationship between them form the basis for analysis in systems theory. The fact that $x(t)$ might represent a voltage or that the system might be an electric motor has little bearing on the theoretical analysis from this point on. This does not mean, however, that practical considerations regarding the physical nature of the problem are unimportant. In real-world problems, physical constraints of the problem at hand and deviations of physical phenomena from idealized mathematical models must be taken into consideration on a case by case basis in parallel to theoretical analysis. Systems theory simply provides a common framework for the analysis of vastly different physical systems.

One detail that needs to be elaborated is the distinction between a discrete-time signal and a digital signal. The time variable is discrete in both signals; that is, both are defined as functions of an integer index. A discrete-time signal consists of samples that are continuous in amplitude. The amplitude of a sample can take any value. On the other hand, samples of a digital signal are quantized and can only take amplitude values from a predefined set. In this text, we will mostly work with discrete-time signals, rather than digital signals, since, on paper, any amplitude can be represented with infinite precision. When we represent signals in a computer, however, they become digital signals. Computers represent numbers with bit streams, and therefore have finite precision.

2.1 ELEMENTARY DISCRETE-TIME SIGNALS

A number of basic discrete-time signals appear very often in digital signal-processing literature and are typically used as building blocks in describing other, more complicated signals.

The *discrete unit-impulse sequence*, also referred to as a *unit-sample sequence* in some textbooks, is defined as

$$\delta(n) = \begin{cases} 1, & \text{if } n = 0 \\ 0, & \text{otherwise.} \end{cases} \quad (2.1)$$

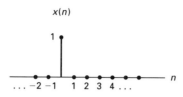

Figure 2.4 Discrete-impulse sequence.

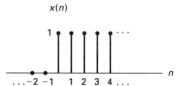

Figure 2.5 Unit-step sequence.

It is important to realize that $\delta(n)$ does not consist of just one sample. Rather, it is an infinite-length sequence in which all samples but one happen to be equal to zero in amplitude. (See Fig. 2.4.)

The *unit-step sequence* is defined as

$$u(n) = \begin{cases} 1, & \text{if } n \geq 0 \\ 0, & \text{otherwise.} \end{cases} \quad (2.2)$$

and is shown in Fig. 2.5. It is possible to express the unit-step sequence in terms of the discrete-impulse sequence as

$$u(n) = \sum_{k=-\infty}^{n} \delta(k). \quad (2.3)$$

Equation (2.3) represents the sum of an infinite number of terms. If the argument, and thus the upper limit of the summation, is nonnegative, the term $\delta(0)$ is among the terms added, and the result is unity. For negative values of the argument, the result is zero.

The *complex exponential sequence* is defined as

$$x(n) = e^{j\omega n} \quad (2.4)$$

where ω is the *angular frequency* in the range from 0 and 2π radians. It can easily be shown that values of ω outside this range can be normalized to values between 0 and 2π radians since increasing or decreasing ω in (2.4) by integer multiples of 2π does not affect the result. Using Euler's formula, it is also possible to express the complex exponential sequence as

$$x(n) = \cos(\omega n) + j\sin(\omega n).$$

A *right-sided exponential sequence* is defined as

$$x(n) = a^n u(n), \quad (2.5)$$

where the parameter a is, in general, a complex constant. The factor $u(n)$ causes $x(n)$ to be zero for $n < 0$. Thus, a functional description of the right-sided exponential sequence is

$$x(n) = \begin{cases} a^n, & \text{if } n \geq 0 \\ 0, & \text{otherwise.} \end{cases}$$

In the case where a is real, $x(n)$ is a decaying function of n if $|a| < 1$ and a growing function of n if $|a| > 1$. If $a = 1$, then $x(n)$ becomes the unit-step sequence. Figure 2.6

Sec. 2.1 Elementary Discrete-time Signals

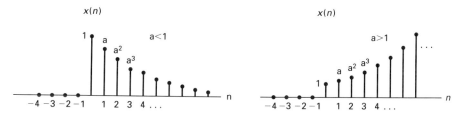

Figure 2.6 Right-sided exponential sequence $x(n) = a^n u(n)$ for real values of a.

illustrates these three cases. If the parameter a is complex, then it is possible to express it in polar complex number form as

$$a = re^{j\theta}.$$

Substituting this into Eq. (2.5) and applying Euler's formula, we obtain

$$x(n) = (re^{j\theta})^n = r^n \cos(\theta n) + jr^n \sin(\theta n).$$

If the magnitude r is less than unity, real and imaginary parts of the right-sided exponential sequence are cosine and sine functions with exponentially decaying amplitudes. In Fig. 2.7, real and imaginary parts of $x(n)$ are shown for parameter values $r = 0.8$ and $\theta = 0.2$ If $r > 1$, then the samples of the sequence $x(n)$ grow without bound.

Exercise 2.1.1

Using the programming language of your choice, develop functions to represent each of the elementary signals mentioned previously.

(a) The functions *Impulse()* and *UnitStep()* should take one integer argument and return the appropriate result in a fashion very similar to the example function *MySignal()* given previously.

(b) The functions *Sine()* and *Cosine()* should take a floating-point argument for the angular frequency ω in addition to the integer argument for n.

(c) The function *ComplexExp()* should also take two arguments for ω and n, respectively. It should make use of the functions *Sine()* and *Cosine()* based on Euler's formula and return the appropriate complex value. Explore the possibility of representing complex numbers in the programming language that you use.

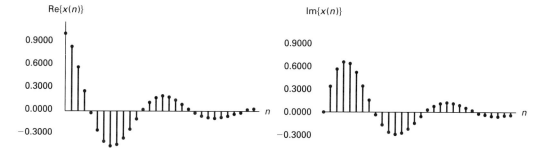

Figure 2.7 Right-sided exponential sequence $x(n) = a^n u(n)$ for complex a.

(d) The function *RSidedExp( )* should take a complex argument for *a* and an integer argument for *n*. It should also return a complex value.

Exercise 2.1.2

The goal of this exercise is to provide an introduction to waveform synthesis capabilities of PC-DSP. Use the *Waveform-generator* function of PC-DSP to generate the following signals. Tabulate and plot each signal in its specified (nonzero) range. Note that all the signals specified next are generated by making the selections *Sequences/Generate-sequence/Waveform-generator* in the menu structure of PC-DSP. Once a signal is generated, it can be graphed on the screen by pressing the F2 key. Alternatively, a table of sample amplitudes can be obtained by pressing the F3 key.

(a) The discrete impulse sequence $\delta(n)$ for $n = 0, \ldots, 99$.
(b) The unit-step sequence $u(n)$ for $n = 0, \ldots, 199$.
(c) The unit-ramp sequence $r(n) = n$ for $n = 0, \ldots, 99$.

Note: The step sequence generated in part (b) is only a finite-length approximation to the unit-step sequence $u(n)$, which is theoretically of infinite length. The same applies to the sequences in parts (b) and (c). Through computer simulation, we are only able to generate finite-length segments of infinite-length signals. This generally does not cause any problems in analysis as long as care is taken to ensure that approximation methods are used properly.

Exercise 2.1.3

(a) Using the waveform generator function of PC-DSP, obtain samples of the complex exponential sequence

$$x(n) = 0.7e^{-j\omega_0 n}u(n)$$

for $\omega_0 = \pi/8$ and for $0 \leq n \leq 99$. Tabulate and graph the resulting sequence. Using Euler's formula, $x(n)$ can be expressed in Cartesian form as

$$x(n) = 0.7 \cos(\omega_0 n)u(n) + j0.7 \sin(\omega_0 n)u(n).$$

Do the real and imaginary parts of the 100-point sequence you created agree with this analytical form?

(b) Graph the magnitude and the phase of $x(n)$. This can be accomplished by pressing *P* while the Cartesian-form graph is on the screen.

2.2 CLASSIFICATION METHODS FOR DISCRETE-TIME SIGNALS

Periodicity

A discrete-time signal $x(n)$ is said to be *periodic* if an integer N can be found such that the equation

$$x(n + N) = x(n) \tag{2.6}$$

is satisfied for all integer values of the index *n*. The parameter N is called the *period* of the discrete-time signal. A signal that does not satisfy this criterion for all integer values of the index *n* is called an *aperiodic* signal. Consider the sinusoidal sequence

$$x(n) = \sin(\omega n), \quad -\infty < n < \infty.$$

Sec. 2.2 Classification Methods for Discrete-time Signals

For $x(n)$ to be periodic, we require

$$\sin[\omega(n + N)] = \sin(\omega n)$$

for all integer values of n and a particular integer value of N. For two sinusoids to be equal, their arguments must either be equal or they must differ by an integer multiple of 2π. Therefore,

$$\omega N = 2\pi k$$

and the period is

$$N = \frac{2\pi k}{\omega} \qquad (2.7)$$

provided that this yields an integer value for N. This leads to an important result: In contrast with a continuous-time sinusoid, which is always periodic, a discrete-time sinusoid may or may not be periodic.

Exercise 2.2.1

In this exercise, discrete-time sinusoids will be generated using PC-DSP, and their periodicity will be examined. Discrete sinusoids can be generated using the same sequence of menu selections as in the previous exercise. Specifically, follow these steps:

(a) Generate 100 samples of a discrete-time sinusoid with the angular frequency $\omega = 0.2\pi$ radians. Tabulate the samples of this sequence, and determine the period by finding the first sample with which the pattern starts repeating. Compute the period using (2.7) and compare. Plot the sequence and observe its general shape. How many cycles of the sinusoid are in one period?

(b) Repeat part (a) with $\omega = 0.15\pi$ radians. Is $x(n)$ still periodic? If it is, what is the period? How many cycles are in one period? How does this correspond to the smallest value of k that yields an integer value for the period N in (2.7)?

(c) Repeat part (a) with $\omega = 0.3$ radians. Is the sequence still periodic? Can you find a value for k that would yield an integer N in (2.7)?

Geometric Series and Closed Forms

In working with summations, it is often possible (and desirable) to obtain a simple closed-form expression to replace a summation. We will have occasion to do this rather frequently in the remainder of the book. Therefore, we will digress slightly here and develop the formulas necessary for this conversion. It is important to understand the derivations that follow so that similar techniques may be employed in a variety of situations.

Consider the infinite-length summation

$$A = \sum_{n=0}^{\infty} \alpha^n. \qquad (2.8)$$

Obviously, if $|\alpha| \geq 1$, then the summation does not converge to a finite value, and thus a closed-form cannot be found. We will only be concerned with the case $|\alpha| < 1$. The first few terms of the summation can be written in open form as

$$A = 1 + \alpha + \alpha^2 + \alpha^3 + \alpha^4 + \ldots . \qquad (2.9)$$

Subtracting 1 from each side of (2.9), we obtain

$$\begin{aligned} A - 1 &= \alpha + \alpha^2 + \alpha^3 + \alpha^4 + \ldots \\ &= \alpha(1 + \alpha + \alpha^2 + \alpha^3 + \ldots) \\ &= \alpha A. \end{aligned} \qquad (2.10)$$

Solving (2.10) for A yields

$$A = \frac{1}{1 - \alpha}. \qquad (2.11)$$

Exercise 2.2.2

Consider the infinite-length geometric series given by (2.8). Provided that α is less than unity in magnitude, the terms of the summation decay rather rapidly and become insignificant for large n. Therefore, the sum of the geometric series can be numerically approximated by adding its first few terms.

(a) For $\alpha = 0.5$, manually compute the sum of the first five terms of the series. How does the result compare to the exact value obtained from the closed-form equation? What is the approximation error?

(b) Repeat part (a) with $\alpha = 0.9$. How does the approximation compare to the closed form result now? Does the approximation error increase or decrease? Why?

(c) Repeat part (a) with $\alpha = 0.9$ and the first ten terms of the summation instead of five. Comment on the result. Does the approximation improve when more terms of the summation are used?

Exercise 2.2.3

In some cases, geometric series summations with a finite number of terms are of interest. Consider the finite-length summation

$$A = \sum_{n=0}^{M} \alpha^n$$

where, in contrast with the summation in (2.8), the upper limit is a finite integer M. Using a procedure similar to the one outlined in Eqs. (2.9) through (2.11), show that the closed-form expression for A is

$$A = \frac{1 - \alpha^{M+1}}{1 - \alpha}. \qquad (2.12)$$

Does this closed form require any restrictions to be placed on the value of α? In particular, how would you handle the situation $\alpha = 1$? Also, is it consistent with the closed form for $M \to \infty$?

Exercise 2.2.4

Derive the closed-form expressions for the following infinite summations. In each case, assume that α is less than unity in magnitude.

(a) $\sum_{n=0}^{\infty} \alpha^n \cos(\omega n)$ (b) $\sum_{n=0}^{\infty} n\alpha^n$ (c) $\sum_{n=0}^{\infty} n(n-1)\alpha^n$

Hint: In part (a), write the cosine term using Euler's formula. In parts (b) and (c), differentiate the geometric series of (2.8) with respect to α, and modify the resulting expression to fit it into the desired form.

Sec. 2.2 Classification Methods for Discrete-time Signals   **19**

Exercise 2.2.5

Using the closed-form formulas developed, evaluate the following summations without computer aid.

(a) $\sum_{n=0}^{\infty} (0.8)^n$ (b) $\sum_{n=0}^{\infty} n(0.8)^n$ (c) $\sum_{n=0}^{10} (0.8)^n$ (d) $\sum_{n=5}^{10} (0.8)^n$

Energy and Power Signals

The *energy* of a discrete-time sequence is defined as

$$E = \sum_{n=-\infty}^{\infty} |x(n)|^2 \quad (2.13)$$

provided that the summation converges; that is, it yields a finite value for E. The signals that have finite energy are said to be *energy signals*. From this definition, we can easily deduce that all finite-length signals (signals with a finite number of nonzero samples) are energy signals. In addition, most infinite-length signals that asymptotically decay to zero for $n \to \pm\infty$ also belong to this category.

The *average power* in a discrete-time signal is defined as

$$P = \lim_{M \to \infty} \frac{1}{2M+1} \sum_{n=-M}^{M} |x(n)|^2, \quad (2.14)$$

where first the energy in the middle $2M + 1$ samples of the signal is computed, and then this value is averaged over the number of samples that contribute to the sum. Finally, the limit is taken over M so that the averaging operation encompasses all signal samples. Signals with infinite energy but finite power are called *power signals*.

The *Signal statistics* function of PC-DSP computes the energy of a sequence along with other values of interest. It is accessed using the menu selections *Operations/Statistics/Signal-statistics*.

Exercise 2.2.6

On paper, roughly sketch the general shape of each of the following signals. Classify each signal as either an energy signal or a power signal.

(a) $x(n) = \delta(n)$ (b) $x(n) = u(n)$ (c) $x(n) = \alpha^n u(n)$

Deterministic and Random Signals

Discrete-time signals can also be characterized as either deterministic or random. A signal is deterministic if it can be expressed in an analytical form. As an example,

$$x(n) = e^{-0.2n} \sin(0.3\pi n) u(n)$$

represents a deterministic sequence since, for any integer value of the index n, it is possible to obtain the amplitude of the corresponding sample from this expression. Random signals, on the other hand, do not have analytical descriptions. Instead, they are described through their statistical (probabilistic) properties. Figure 2.8 shows an example of a random sequence.

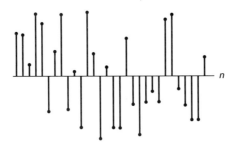

Figure 2.8 Random sequence.

Exercise 2.2.7

PC-DSP contains functions for generating *pseudorandom* discrete-time signals with approximately *Gaussian* or *uniform* probability distribution functions. These functions can be accessed through menu choices *Sequences/Generate-sequence/Random-sequences*.

(a) Generate and plot a 1000-sample random sequence the samples of which come from a Gaussian distribution with mean $\mu = 0$ and standard deviation $\sigma = 1$.

(b) Generate and plot a 1000-sample random sequence the samples of which come from a uniform distribution with sample amplitudes between 0 and 2.

(c) A *histogram* is a function that provides information about the characteristics of a random signal. The amplitude range of the random signal is divided into a number of subintervals (bins). For each sample of the random signal, the occurrence count of the corresponding bin is incremented. The histogram is then constructed as a discrete sequence of these counts. Using the histogram function of PC-DSP, compute and plot histograms of the random signals generated in parts (a) and (b). For the Gaussian sequence, use 30 bins between -3 and $+3$. For the uniform sequence, use 20 bins between 0 and 2. The histogram function is accessed through menu choices *Operations/Statistics/Histogram*.

2.3 ELEMENTARY OPERATIONS ON DISCRETE-TIME SIGNALS

In this section, we will define a number of arithmetic operations for manipulating discrete-time sequences. These operations allow sequences to be analytically expressed in terms of simpler sequences.

Addition of sequences. Two sequences are added by adding the values of their corresponding samples for each value of the index n. Thus, the sum of two sequences is a new sequence each sample of which is equal to the sum of the corresponding samples in the two sequences being added; that is,

$$y(n) = x_1(n) + x_2(n).$$

Multiplication of sequences. The product of two sequences is a new sequence each sample of which is equal to the product of the corresponding samples in the two sequences being multiplied; that is,

Sec. 2.3 Elementary Operations on Discrete-time Signals

$$y(n) = x_1(n)x_2(n).$$

Scalar addition. Adding a constant scalar offset to a sequence means adding the scalar to each sample of the sequence; that is,

$$y(n) = x(n) + \alpha.$$

Multiplication by a constant. Multiplying a sequence by a constant scalar means multiplying each sample of the sequence with the constant; thus,

$$y(n) = \alpha x(n).$$

Time shifting. Replacing the integer argument n of a sequence by $n - k$ causes the sequence to be delayed (shifted to the right) by k samples if k is positive. If k is negative, then the shift is to the left, corresponding to an advance.

To understand the mechanism that causes the shift, consider again the programming example given at the beginning of this chapter. Recall that the function *MySignal(n)* returns the values 1.7, -0.6, and 2.3 when the argument n is equal to 0, 1, or 2, respectively. Let's define a new function *NewSignal(n)* as follows:

```
float NewSignal (int Argument)
{
    return MySignal (Argument-5);
}
```

The relationship between the two signals represented by these two functions is

$$NewSignal(n) = MySignal(n - 5)$$

If, for example, the function *NewSignal()* is called with the index value 6, it calls the function *MySignal()* with the index value 1, thus causing the value -0.6 to be returned. The function *NewSignal()* returns 1.7, -0.6, and 2.3 when its argument is 5, 6, and 7, respectively. It is thus a five-sample delayed version of *MySignal()*.

Time reversal. Replacing the integer argument n of a sequence by $-n$ causes the sequence to be reversed in time, that is, flipped around the time origin $n = 0$. It is possible to justify this by using a similar logic as in the previous time-shifting example.

Exercise 2.3.1

Roughly sketch each of the following sequences by hand. Afterward use PC-DSP to generate and plot each sequence within the specified range of the index. Note that $r(n)$ denotes the unit-ramp sequence; that is,

$$r(n) = n\, u(n).$$

In PC-DSP, arithmetic operations can be performed by going through the menu selections *Operations/Arithmetic-operations* and then selecting the appropriate operator from the pull-down menu. In time-shifting sequences, left shifts can be achieved using negative values for the delay.

(a) $x(n) = u(n) - u(n - 20),$ $\qquad n = 0, \ldots, 99$

(b) $x(n) = e^{-0.7n}[u(n) - u(n - 20)]$, $\qquad n = 0, \ldots, 99$

(c) $x(n) = r(n)[u(n - 5) - u(n - 25)]$, $\qquad n = 0, \ldots, 30$

(d) $x(n) = r(n + 5) - r(n - 5)$, $\qquad n = 0, \ldots, 30$

(e) $x(n) = \sin(0.3\pi n)u(n)$, $\qquad n = 0, \ldots, 99$

(f) $x(n) = \sin(1.3\pi n)u(n)$, $\qquad n = 0, \ldots, 99$

Note: In part (d), applying the *Shift* function to the first ramp sequence with a delay value of -5 samples will cause the first sample to appear at $n = -5$. To remove the undesired samples for $n = -5, \ldots, -1$, use the *Copy* function accessed by menu selections *Sequences/Sequence-editing/Copy-sequence*. Specify the copy index range from 0 to 30.

Exercise 2.3.2

In this exercise, programming constructs will be developed for generating the discrete-time signals in Exercise 2.3.1. As an example, the sequence $x(n)$ in part (a) of that exercise can be implemented in C language as follows:

```
float Xa(int Argument)
{
    return UnitStep(Argument)-UnitStep(Argument-20);
}
```

This implementation uses the function *UnitStep()* developed in Exercise 2.1.1. The extension of this idea to any other programming language should be obvious. Using the programming language of your choice, develop functions to generate samples of the sequences (b) through (f) in Exercise 2.3.1. Whenever possible, use the functions developed in Exercise 2.1.1 as building blocks. Use each function in a loop construct for the specified index range to compute the corresponding samples.

Exercise 2.3.3

In this exercise, we will generate a symmetric exponential sequence

$$x(n) = e^{-0.05|n|}$$

for the range of the index $n = -99, \ldots, 99$.

(a) Using the waveform generator in PC-DSP, generate a one-sided exponential

$$x_1(n) = e^{-0.05n}u(n), \qquad n = 0, \ldots, 99.$$

(b) Generate a time-reversed version of $x_1(n)$ using the menu selections *Operations/Arithmetic-operations/Flip-sequence*. At this point both $x_1(n)$ and its time-reversed version contain a sample at $n = 0$. Remove the sample at $n = 0$ in the time-reversed exponential using the menu selections *Sequences/Sequence-editing/Copy-sequence*. Specify the copy index range from -1 to -99.

(c) Add the sequences of parts (a) and (b) to obtain the desired sequence. Tabulate and plot the result.

Exercise 2.3.4

The goal of this exercise is to generate a periodic triangular sequence using the *Ramp* function of PC-DSP along with functions for time reversal and periodic extension.

(a) The first task is to generate a single triangle that is centered around the origin and that extends to $n = \pm 20$ in either direction. Use the waveform generator of PC-DSP to generate a 21-

Sec. 2.3 Elementary Operations on Discrete-time Signals

sample ramp with an initial amplitude of 1 and a slope of $-\frac{1}{20}$. Tabulate and plot the result to check its accuracy. This should give us the right side of the triangle.

(b) Obtain the left side of the triangle by time-reversing the right side and then removing the sample at $n = 0$. Tabulate and plot the left side. Add the two sides to complete the triangle.

(c) To make a periodic sequence out of the single triangle, use the menu selections *Sequences/Sequence-editing/Make-periodic* and specify the range of the index from $n = -99$ to $n = 99$.

Exercise 2.3.5

In this exercise, amplitude modulation of continuous-time signals will be simulated using discrete-time sequences. In communication systems, it is often necessary to transmit a message signal by means of a high-frequency sinusoid (called *the carrier*). One method of achieving this is to vary (modulate) the amplitude of the carrier in proportion with the message signal. An amplitude-modulated (AM) signal has the general form

$$x_a(t) = A_c[1 + \mu m(t)] \cos(2\pi f_c t),$$

where $m(t)$ represents a *message signal*. The parameter f_c is the carrier frequency, and μ is called the *modulation index*. We will consider the special case in which the message signal is also a sinusoid; that is,

$$x_a(t) = A_c[1 + \mu \sin(2\pi f_m t)] \cos(2\pi f_c t)$$

Figure 2.9 shows this signal for $f_c = 1$ kHz, $f_m = 100$ Hz, $\mu = 0.7$, and $A_c = 1$. A discrete sequence $x(n)$ can be obtained by evaluating this expression at time instants that are integer multiples of a predefined time interval T; that is, $t = nT$.

$$x(n) = A_c[1 + \mu \sin(2\pi f_m nT)] \cos(2\pi f_c nT).$$

Using $T = 10^{-5}$, generate 1024 samples of the sequence $x(n)$. Use the values of f_c, f_m, μ, and A_c given previously. Graph the sequence. You should obtain the plot shown in Fig. 2.9. *Note*: To observe this sequence as an approximation to the continuous-time signal, graph it in continuous mode. This is achieved by pressing C while the graph is displayed on the screen.

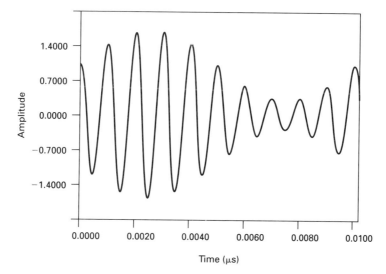

Figure 2.9 Amplitude-modulated signal of Exercise 2.3.5.

Exercise 2.3.6

In this exercise, we will consider the problem of generating samples of the sequence

$$h(n) = (-0.9)^n u(n)$$

for the range $n = 0, \ldots, 99$. Even though PC-DSP has a built-in function for generating a right-sided exponential sequence directly, we will ignore it for the time being and use the functions *Exponential* and *Logarithm* for this purpose. It can be easily shown that

$$a^n = e^{n \ln(a)}.$$

Use this property to generate the sequence $h(n)$ in five steps:

(a) Generate a 100-sample sequence $h_1(n)$ with all samples equal to -0.9. The easiest method is to use the *Step* function of the waveform generator with an amplitude of -0.9. Recall that the waveform generator is accessed through *Sequences/Generate-sequence/Waveform-generator*.

(b) Compute $h_2(n)$ as the natural logarithm of the sequence $h_1(n)$ using the menu selections *Operations/Nonlinear-operations/Logarithm*.

(c) Generate $h_3(n)$ as a ramp sequence made of integer valued samples $0, 1, \ldots, 99$. The easiest way to do this is to use the *Ramp* function of the waveform generator.

(d) Obtain $h_4(n)$ by multiplying the sequences $h_2(n)$ and $h_3(n)$ sample by sample. The access sequence is *Operations/Arithmetic-operations/Multiply-sequences*.

(e) Finally, compute $h(n)$ by applying the *Exponential* function to $h_4(n)$. This function is accessed with *Operations/Nonlinear-operations/Exponential*.

Note: The resulting sequence $h(n)$ will be complex, but tabulation of sample values should reveal that imaginary parts are insignificant and are basically due to machine round-off errors. It is possible to remove imaginary parts with menu selections *Operations/Arithmetic-operations/Real-part*.

One interesting point in Exercise 2.3.6 is the natural logarithm operation in step (b). We computed the natural logarithm of the sequence $h_1(n)$ which had negative samples. The obvious question is how to compute the logarithm of a negative number. The answer lies in the fact that all but a few PC-DSP functions are designed to work with complex sequences. Definitions of most traditional functions can be extended to cover complex numbers. For example, the natural logarithm of a complex number in polar form $re^{j\theta}$ can be computed as

$$\ln(re^{j\theta}) = \ln(r) + \ln(e^{j\theta})$$
$$= \ln(r) + j\theta.$$

Since any negative number can also be written as a complex number in polar form, it is possible to compute the logarithm of negative numbers if we are willing to accept complex results. (In this case, we are.) Using this philosophy, it can be shown that the natural logarithm of -0.9 is $-0.105361 + j\pi$.

Exercise 2.3.7

Generate the signal described in the previous exercise, this time using the built-in function for obtaining right-sided exponential sequences. Use the menu selections *Sequences/Generate-sequence/Waveform-generator*, and specify a right-sided exponential. Tabulate and graph the resulting sequence. Compare it to the sequence obtained in Exercise 2.3.6.

Sec. 2.3 Elementary Operations on Discrete-time Signals

Exercise 2.3.8

On paper, roughly sketch the sequence

$$p(n) = \frac{\sin(0.1n)}{0.1n}.$$

Afterward, use PC-DSP to generate samples of this sequence for $n = 0, \ldots, 99$. Although there is a faster way of generating a sequence of this type directly, we will take the longer route and do it in several steps. Let us make the following definitions:

$$r(n) = 0.1n, \qquad n = 0, \ldots, 99,$$
$$s(n) = \sin[r(n)], \qquad n = 0, \ldots, 99,$$
$$q(n) = \frac{1}{r(n)} \qquad n = 0, \ldots, 99.$$

The desired sequence $p(n)$ can then be expressed as the product of $q(n)$ and $s(n)$. We already know how to generate $r(n)$ using the *Ramp* function of the waveform generator. Once $r(n)$ is obtained, $s(n)$ and $q(n)$ can be produced using *Sine* and *Reciprocate* functions, respectively. Both of these functions are accessed by going through menu selections *Operations/Nonlinear-operations*.

One potential problem needs to be addressed here. The first sample of the sequence $r(n)$ is zero, and an attempt to reciprocate $r(n)$ would result in a floating-point overflow. (You may try this if you wish. The program does not crash; it just displays an error message and refuses to do the operation that causes the overflow.) A practical solution to the overflow problem is to add a very small positive number to each sample in the sequence $r(n)$ before using it for computing $s(n)$ and $q(n)$. For this particular problem, try an offset $\epsilon = 10^{-5}$. Is the sample at $n = 0$ correct? Would the same value be obtained if the *Sine* function of the waveform generator were used instead of the *Sine* function of the *Nonlinear-operations* menu?

Exercise 2.3.9

Generate the signal described in the previous exercise using the formula-entry method. Use the menu selections *Operations/Generate-sequence/Formula-entry*, and specify the desired range. The index value $n = 0$ is still critical and causes a division-by-zero error unless caution is taken. To prevent this error, add a small fractional value to the index; that is, enter the formula as

$$\sin(0.1*(n + 0.00001))/0.1*(n + 0.00001)$$

This prevents the division by zero at $n = 0$ and should not affect the accuracy of other samples significantly. Tabulate the generated sequence simultaneously with the sequence obtained in Exercise 2.3.8, and compare sample values.

Any arbitrary sequence $x(n)$ can be decomposed into a set of scaled and shifted unit-impulse sequences. Consider the sequence $x_k(n)$ obtained by multiplying the shifted unit-impulse sequence $\delta(n - k)$ with the kth sample of $x(n)$; that is,

$$x_k(n) = x(k)\delta(n - k).$$

The only nonzero sample of $x_k(n)$ occurs at $n = k$, and its amplitude is equal to the amplitude of the kth sample of $x(n)$. (See Fig. 2.10.) By repeating this procedure for all integer values of k and adding the results, $x(n)$ can be written as

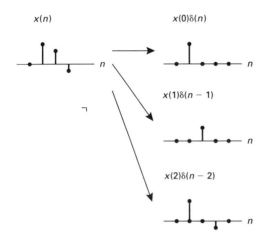

Figure 2.10 Decomposition of a sequence into scaled and shifted impulse sequences.

$$x(n) = \sum_{k=-\infty}^{\infty} x_k(n) \quad (2.15)$$

$$= \sum_{k=-\infty}^{\infty} x(k)\delta(n-k).$$

This decomposition will be very useful in the next chapter when we consider linear systems and superposition.

2.4 SYMMETRY PROPERTIES OF DISCRETE-TIME SIGNALS

Some discrete-time signals exhibit symmetry properties that could be exploited to simplify their analysis. Consider a real sequence $x(n)$. An even sequence is one that satisfies the condition

$$x(-n) = x(n) \quad (2.16)$$

for all integer values of n. In contrast, an odd sequence is characterized by

$$x(-n) = -x(n). \quad (2.17)$$

A particular sequence does not have to be either even or odd, but any sequence can be written as a sum of an even sequence and an odd sequence:

$$x(n) = x_e(n) + x_o(n), \quad (2.18)$$

where the even component is computed as

$$x_e(n) = \frac{1}{2}[x(n) + x(-n)] \quad (2.19)$$

and the odd component is

$$x_o(n) = \frac{1}{2}[x(n) - x(-n)]. \quad (2.20)$$

Sec. 2.4 Symmetry Properties of Discrete-time Signals

Exercise 2.4.1

Verify that, for any arbitrary sequence $x(n)$, the sequences $x_e(n)$ and $x_o(n)$ as defined in Eqs. (2.19) and (2.20) satisfy even and odd sequence definitions, respectively, and that their sum is equal to $x(n)$.

Exercise 2.4.2

Generate a 100-sample real sequence $x(n)$ with arbitrary sample amplitudes. (Use the random sequence generator accessed through *Sequences/Generate-sequence/Random-sequences*.) Compute its even and odd components by direct application of (2.19) and (2.20). Even though PC-DSP has built-in functions for computing the even and odd components of a signal, we will take a longer route and use the *Add*, *Subtract*, *Flip* and *Multiply-by-constant* functions of the *Arithmetic-operations* submenu. Once the even and the odd components are obtained, verify (2.18) by adding them and comparing the result to $x(n)$.

Exercise 2.4.3

Consider the sequence $x(n)$ that was generated in part (d) of Exercise 2.3.1, which will be repeated here for convenience:

$$x(n) = r(n + 5) - r(n - 5), \qquad n = 0, \ldots, 30$$

(a) By hand, roughly sketch the even and odd components of this sequence using Eqs. (2.19) and (2.20). How many nonzero samples do the even and odd components have?

(b) Compute and plot the even and odd components of $x(n)$ using PC-DSP. Note that these components can be obtained using the menu selections *Operations/Arithmetic-operations/Even-part* and *Operations/Arithmetic-operations/Odd-part*, respectively. Compare the results to your hand sketches.

(c) Verify (2.18) for the sequences you obtained by adding them and comparing the sum to the original sequence $x(n)$.

The definitions of the even and odd symmetry properties given could also be applied to complex sequences. It is possible to decompose a complex sequence into its even and odd components by treating its real and imaginary parts separately and using (2.18) on each. However, *conjugate symmetry* properties are often more useful for problems involving frequency-domain transforms of complex sequences. A sequence is said to be *conjugate symmetric* if it satisfies the condition

$$x(-n) = x^*(n) \tag{2.21}$$

for all n. Note that the asterisk (*) denotes complex conjugate. In contrast, a *conjugate-antisymmetric* sequence is one that satisfies the condition

$$x(-n) = -x^*(n). \tag{2.22}$$

It is possible to write an arbitrary complex sequence as the sum of two components as

$$x(n) = x_E(n) + x_O(n), \tag{2.23}$$

where the conjugate-symmetric and the conjugate-antisymmetric components are computed by

$$x_E(n) = \frac{1}{2}[x(n) + x^*(-n)], \qquad (2.24)$$

$$x_o(n) = \frac{1}{2}[x(n) - x^*(-n)]. \qquad (2.25)$$

As a special case, if $x(n)$ is real, then Eqs. (2.18) through (2.20) are equivalent to Eqs. (2.23) through (2.25).

Exercise 2.4.4

(a) Use PC-DSP to generate a 100-sample complex sequence ($n = 0, \ldots, 99$) with arbitrary sample values. *Suggestion:* Generate two real pseudorandom sequences using the menu selections *Sequences/Generate-sequence/Random-sequences*. Make a complex sequence out of these two real sequences using the menu selections *Sequences/Sequence-editing/Make-complex*.

(b) Compute conjugate-symmetric and conjugate-antisymmetric components of the complex sequence obtained in part (a). Use elementary signal operations, and compute the required components by direct application of (2.24) and (2.25).

(c) Compute conjugate-symmetric and conjugate-antisymmetric components of the complex sequence obtained in part (a), this time using the menu selections *Operations/Arithmetic-operations/Even-part* and *Operations/Arithmetic-operations/Odd-part*, respectively. In each case, check the option box with the text *Conjugate* to indicate that the conjugate symmetry definitions of Eqs. (2.24) and (2.25) are to be used instead of (2.19) and (2.20). Also, make sure that the option box with the title *Modulo-N* is left unchecked.

(d) Tabulate the sequences obtained. For the conjugate-symmetric component, compare the sample $x_E(10)$ to the sample $x_E(-10)$. Repeat with the conjugate-antisymmetric component.

2.5 DISCRETE-TIME SINUSOIDS REVISITED

In this section, we will develop the relationship between continuous-time sinusoids and their discrete-time counterparts. Consider a continuous-time sinusoid in the form

$$\begin{aligned} x_a(t) &= A\cos(\Omega_0 t) \\ &= A\cos(2\pi f_0 t), \end{aligned} \qquad (2.26)$$

where Ω_0 is the frequency in radians per second and f_0 is in Hertz. A discrete-time sinusoid can be obtained by evaluating $x_a(t)$ at integer multiples of a specified increment of time; that is,

$$\begin{aligned} x(n) &= x_a(nT) \\ &= A\cos(2\pi f_0 nT). \end{aligned} \qquad (2.27)$$

This corresponds to sampling the continuous-time sinusoid every T seconds. The parameter T is therefore referred to as the *sampling interval*. The *sampling frequency*, defined as the number of samples taken from $x_a(t)$ in unit time, is

$$f_s = \frac{1}{T}. \qquad (2.28)$$

Sec. 2.5 Discrete-time Sinusoids Revisited

At this point, we will define the *normalized frequency* of the discrete-time sinusoid $x(n)$ as

$$F_0 = \frac{f_0}{f_s} = f_0 T. \tag{2.29}$$

Obviously, the normalized frequency is a dimensionless quantity. With this definition, $x(n)$ can be written as

$$x(n) = A \cos(2\pi F_0 n). \tag{2.30}$$

This last expression is identical in form to the expression for $x_a(t)$ given by (2.26). Its significance lies in the fact that the use of the normalized frequency concept makes it possible to manipulate a discrete-time sinusoidal sequence without knowing the sampling interval or the sampling frequency that might have been used in obtaining it. The angular frequency (in radians) of the discrete-time sinusoid is

$$\omega_0 = 2\pi F_0, \tag{2.31}$$

and (2.30) can also be written in the form

$$x(n) = A \cos(\omega_0 n). \tag{2.32}$$

It is interesting to note that two discrete-time sinusoids with angular frequencies that differ by an integer multiple of 2π are indistinguishable from each other; that is,

$$A \cos(\omega_0 n) = A \cos([\omega_0 + 2\pi k]n). \tag{2.33}$$

The direct result of (2.33) is that a discrete-time sinusoid is only unique in a 2π radian range of the angular frequency ω_0. If we restrict ω_0 to be in the range $-\pi \leq \omega_0 < \pi$, then the normalized frequency F_0 will be in the range $-0.5 \leq F_0 < 0.5$.

Exercise 2.5.1

(a) A 100-Hz continuous sinusoid is sampled with a sampling rate of $f_s = 250$ Hz to obtain a sequence $x(n)$. Using the waveform generator function of PC-DSP, generate the first 100 samples of this sequence.

(b) A 350-Hz continuous sinusoid is sampled with the same sampling rate $f_s = 250$ Hz to obtain another sequence $w(n)$. Again using PC-DSP, generate the first 100 samples of this sequence. Compare the two sequences $x(n)$ and $w(n)$. Comment on the results.

(c) You should conclude that the same set of samples is obtained for both sequences $x(n)$ and $w(n)$, even though the continuous signals that led to these sequences were at different frequencies. This is known as the *aliasing* effect and will be explored further in Chapter 5. For the moment it will suffice to say that, under the given circumstances, the sinusoid at 350 Hz acts as an *alias* for the sinusoid at 100 Hz. Figure 2.11 illustrates this. Dashed lines represent the continuous sinusoids, and the circles represent the time instants at which the samples are taken. Show that 350 Hz is not the only alias of 100 Hz for the given sampling rate, and other aliases exist. Find three other frequencies that are also aliases of 100 Hz. For each, generate the first 100 samples of the sampled sequence, and compare to $x(n)$ and $w(n)$.

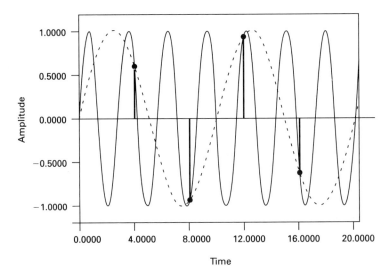

Figure 2.11 Aliasing effect in sampling continuous sinusoids.

3

Discrete-time Systems

> Learning without thought is labor lost;
> thought without learning is perilous.
> —Confucius

In Chapter 2, we reviewed the definitions and the basic properties of discrete-time signals. Most digital signal-processing applications involve the manipulation of discrete-time signals through discrete-time systems. In this sense, the mathematical model of a discrete-time system is a formula or an algorithm that transforms the input signal $x(n)$ into the output signal $y(n)$. Figure 3.1 illustrates this relationship between the input and the output signals using a block-diagram approach. Even though a single-input, single-output system is shown in the figure, it is possible to extend the same idea to systems with multiple inputs and/or outputs. The actual system that implements this mathematical model would consist of digital computer hardware on which the algorithm is programmed.

The mathematical relationship between $x(n)$ and $y(n)$ can be expressed as

$$y(n) = T[x(n)] \qquad (3.1)$$

where $T[..]$ is the mathematical transformation function that represents the action of the system. Signal-system interaction problems generally fall into one of two basic categories: (1) analysis problems and (2) identification or design problems.

In analysis problems, a complete description of the system is known; that is, the transformation $T[..]$ is available in some form. Simply stated, the goal is to determine the

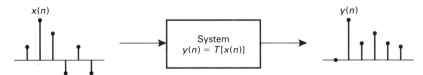

Figure 3.1 Block diagram representation of a discrete-time system.

output signal y(n) for a given input signal x(n). In a broader sense, however, our main interest is in determining the general behavior characteristics, that is, the "attitude" of the system toward certain types of input signals, rather than its response to one particular input signal.

In identification and design problems, input and output signals are known, yet the system is unknown. The problem is to determine the transformation $T[..]$ that would satisfy (3.1) either exactly or approximately. One possible scenario might be an identification problem in which a physical system exists, but its characteristics are unknown. The system's response to certain types of test signals could be observed and/or measured, and a mathematical model could be found to approximately match the behavior of the system. Once a potential mathematical model is found, the identification problem might revert to an analysis problem; the behavior of the model could be analyzed and checked against the actual system either to validate or refute the model. Another possible scenario might be a design problem in which a system is needed to provide a certain type of relationship between the input and the output signals. If a suitable transformation $T[..]$ is found, the system could be built.

In this chapter, fundamental principles of discrete-time systems will be reviewed. Time-and frequency-domain methods of describing and/or analyzing discrete-time systems will be introduced.

3.1 LINEAR AND TIME-INVARIANT SYSTEMS

As stated in (3.1), a discrete-time system can be modeled by means of a transformation formula or algorithm $T[..]$. It is also customary to categorize systems based on the properties of this transformation.

Linearity

A discrete-time system is said to be linear if it satisfies both of the following conditions:

(a) $T[x_1(n) + x_2(n)] = T[x_1(n)] + T[x_2(n)]$, for all $x_1(n)$, $x_2(n)$, and n; (3.2)

(b) $T[\alpha x(n)] = \alpha T[x(n)]$, for all $x(n)$, α, and n. (3.3)

The first condition requires that the response of the system to the sum of two signals be equal to the sum of the responses to individual signals. It can be shown that, if this condition is satisfied for any two signals $x_1(n)$ and $x_2(n)$, then it is also satisfied for any arbitrary number of signals; that is,

Sec. 3.1 Linear and Time-invariant Systems

$$T[x_1(n) + x_2(n) + x_3(n) + \ldots] = T[x_1(n)] + T[x_2(n)] + T[x_3(n)] + \ldots$$

or, in more compact form

$$T[\sum_i x_i(n)] = \sum_i T[x_i(n)]. \quad (3.4)$$

The second condition of linearity requires that any scaling of the input signal by a constant scale factor result in the scaling of the output signal by the same scale factor. Equations (3.2) and (3.3) can be expressed together as

$$T[\alpha x_1(n) + \beta x_2(n)] = \alpha T[x_1(n)] + \beta T[x_2(n)], \quad (3.5)$$

where α and β are arbitrary constants. This form of the linearity definition will be important later in this chapter.

Exercise 3.1.1

A number of discrete-time systems are described next by means of the transformation $T[..]$. By checking the validity of (3.2) and (3.3) for each system, determine which systems are linear and which systems are not. Computer use is not necessary for this exercise.

(a) $y(n) = x(n) + 2$
(b) $y(n) = x(n) + x(n-1)$
(c) $y(n) = 3x(n)$
(d) $y(n) = [x(n)]^2$
(e) $y(n) = a^{x(n)}$
(f) $y(n) = n\, x(n)$

Exercise 3.1.2

In this exercise, we will test the linearity of a system with unknown characteristics. The system under consideration has been programmed as a PC-DSP macro and is on your distribution disks under the file name EX3-1-2.MAC. It can be executed by selecting *Macros* from the top menu bar, then selecting *ex3-1-2*, and following the prompts. (*Note:* A *macro* is a user-defined extension of PC-DSP. Information on how you can develop macros to implement additional functions using the PC-DSP macro compiler can be found in the on-line help system.)

(a) Generate two arbitrary signals $x_1(n)$ and $x_2(n)$. Keep each signal to 25 samples. As an example, $x_1(n)$ may be a decaying exponential, and $x_2(n)$ may be a random signal with a Gaussian distribution.
(b) Using the two signals generated in part (a), generate the signals $x_1(n) + x_2(n)$, $3x_1(n)$, and $-1.5x_2(n)$.
(c) Test the linearity of the system by testing Eqs. (3.2) and (3.3) on it with the signals generated in part (a). Obtain the following output signals:

$$y_1(n) = T[x_1(n)]$$
$$y_2(n) = T[x_2(n)]$$
$$y_3(n) = T[x_1(n) + x_2(n)]$$
$$y_4(n) = T[3x_1(n)]$$
$$y_5(n) = T[-1.5x_2(n)]$$

Tabulate and plot the output signals in each case. Based on your results, does the system seem to be linear? Note that it is not possible to prove linearity of a system by simply testing it with a few specific signals. Your conclusion should be one of the following: (1) the system *may* be linear, or (2) the system is definitely *not* linear.

Exercise 3.1.3

Repeat the steps of the previous exercise, this time working with the system given by the PC-DSP macro EX3-1-3.MAC. This file is also on your distribution disks.

Exercise 3.1.4

In this exercise, we will test the linearity of each system described in Exercise 3.1.1. To do this, we will use the test signals $x_1(n)$ and $x_2(n)$ that were generated in Exercise 3.1.2. Using these test signals, find the following output signals for each system:

$$y_1(n) = T[x_1(n)]$$
$$y_2(n) = T[x_2(n)]$$
$$y_3(n) = T[x_1(n) + x_2(n)]$$
$$y_4(n) = T[3x_1(n)]$$
$$y_5(n) = T[-1.5x_2(n)]$$

For example, the system in part (a) of Exercise 3.1.1 was defined with

$$y(n) = x(n) + 2,$$

and the corresponding output signals would be

$$y_1(n) = x_1(n) + 2,$$
$$y_2(n) = x_2(n) + 2,$$
$$y_3(n) = x_1(n) + x_2(n) + 2,$$
$$y_4(n) = 3x_1(n) + 2,$$
$$y_5(n) = -1.5x_2(n) + 2.$$

These signals can be obtained using the elementary signal operations in PC-DSP. For parts (a) through (f) of Exercise 3.1.1, obtain these output signals. For each system, tabulate and graph the output signals, and comment on the linearity of the system. Note that part (e) may be somewhat tricky.

Time Invariance

A system is said to be time invariant if a delay in the input signal results in the output signal being delayed by the same amount, but otherwise unchanged. Mathematically, if

$$y(n) = T[x(n)],$$

then

$$y(n - k) = T[x(n - k)], \quad \text{for all } x(n) \text{ and all } k. \tag{3.6}$$

Exercise 3.1.5

A number of discrete-time systems are described next by means of the transformation $T[..]$. Determine which of the systems given are time invariant. Computer use is not necessary for this exercise.

(a) $y(n) = x(n) + 2$
(b) $y(n) = x(n) + x(n - 1)$
(c) $y(n) = [x(n)]^2$
(d) $y(n) = n\, x(n)$
(e) $y(n) = x(|n|)$
(f) $y(n) = x(n) + x(-n)$

Exercise 3.1.6

In this exercise, we will test the time-invariance property of a system with unknown characteristics. The system under consideration has been programmed as a PC-DSP macro and is on your distribution disks under the file name EX3-1-6.MAC.

(a) Generate an arbitrary signal $x(n)$ with 25 samples. You may use a sinusoid, a decaying exponential, or a random signal.

(b) Apply the signal $x(n)$ to the unknown system. Tabulate and plot the output signal.

(c) Delay $x(n)$ by five samples to obtain $x(n-5)$ and apply it to the system. Compare the output signal to that obtained in part (b).

(d) Apply $x(n-10)$ to the system, and compare the output signal to those obtained in parts (b) and (c).

(e) Based on your results, does the system seem to be time invariant? Note that it is not possible to prove time invariance based on a few tests with a particular test signal, and similar considerations apply as in Exercise 3.1.1.

Exercise 3.1.7

Repeat the steps of the previous exercise, this time working with the system given by PC-DSP macro EX3-1-7.MAC. This file is also on your distribution disks.

3.2 SIGNAL-SYSTEM INTERACTION

Consider again the system model shown in Fig. 3.1 and the transformation relationship between the input and the output signals given by Eq. (3.1). A system is said to be *memoryless* if, for a given index value n_1, the output sample $y(n_1)$ depends only on the present input sample $x(n_1)$. As an example, a system with the description

$$y(n) = x(n) + \sin[x(n)]$$

is memoryless. In signal-processing applications, memoryless systems are of limited usefulness, and most systems have memory, meaning that each output sample depends on present as well as past values of the input and/or output signals.

We established in Section 2.3 that any arbitrary sequence $x(n)$ can be written as a sum of shifted and scaled discrete-impulse sequences. Equation (2.15) of that section is repeated here for convenience:

$$x(n) = \sum_{k=-\infty}^{\infty} x(k)\delta(n-k). \tag{3.7}$$

Substituting (3.7) into (3.1), we obtain

$$y(n) = T[\sum_{k=-\infty}^{\infty} x(k)\delta(n-k)], \tag{3.8}$$

which, obviously, is not a very useful equation for determining the output signal $y(n)$, unless the transformation $T[..]$ is a simple memoryless transformation.

If the system is linear, the response of the system to a sum of input signals is equal to the sum of responses to individual signals as shown in (3.4). Therefore, (3.8) can be written in the alternative form

$$y(n) = \sum_{k=-\infty}^{\infty} T[x(k)\delta(n-k)]. \qquad (3.9)$$

Using the second condition of linearity given by (3.3), we obtain

$$y(n) = \sum_{k=-\infty}^{\infty} x(k)T[\delta(n-k)], \qquad (3.10)$$

which suggests that, for a linear system, the response to an arbitrary input signal $x(n)$ can be uniquely determined provided that the response of the system to each shift of the discrete-impulse sequence is known. Equation (3.10) still is not very useful for solving signal-system interaction problems since it requires the knowledge of signals $T[\delta(n)]$, $T[\delta(n-1)], \ldots, T[\delta(n-k)]$ for all integer values of k.

Let $h(n, k)$ be defined as

$$h(n, k) = T[\delta(n-k)]. \qquad (3.11)$$

Thus, (3.10) can be written as

$$y(n) = \sum_{k=-\infty}^{\infty} x(k)h(n, k). \qquad (3.12)$$

Let

$$h(n) = h(n, 0) = T[\delta(n)]. \qquad (3.13)$$

We will refer to $h(n)$ as the impulse response of the system. If, in addition to being linear, the system is also time invariant, then the response to $\delta(n-k)$ is simply a shifted version of the response to $\delta(n)$; that is,

$$h(n, k) = T[\delta(n-k)] = h(n-k) \qquad (3.14)$$

and (3.12) can be simplified to

$$y(n) = \sum_{k=-\infty}^{\infty} x(k)h(n-k), \qquad (3.15)$$

indicating that the knowledge of $h(n)$ is sufficient for determining the response of the system to any arbitrary input signal $x(n)$. It is important to remember that this result is only valid for systems that are both linear and time invariant. Because of the simplifications in analysis, physical systems are modeled with linear and time-invariant mathematical models whenever possible.

Through a change of variables in (3.15), it can be shown that the roles of $x(n)$ and $h(n)$ are interchangeable in the convolution equation. Thus, an alternative form of (3.15) is

$$y(n) = \sum_{k=-\infty}^{\infty} x(n-k)h(k). \qquad (3.16)$$

Sec. 3.2 Signal-system Interaction

Exercise 3.2.1

(a) Using the programming language of your choice, write a routine to evaluate the convolution of two sequences by direct implementation of (3.15). Your routine should take three integer arguments, one for the index *n*, one for the lower limit to be used for the summation in (3.15), and one for the upper limit of the same summation. It should compute and return the *n*th sample of the convolution result. As inputs, it should use the routines *SignalH()* and *SignalX()* written to represent sequences *x(n)* and *h(n)*. An example will be given using the C programming language:

```
float Convolve(int n, int lower, int upper)
{
  float Sum;
  int k;
  Sum = 0.0;
  for (k = lower; k <= upper; k++)
    Sum = Sum + SignalX(k)*SignalH(n - k)
  return (Sum);
}
```

(b) Write the routines *SignalX()* and *SignalH()* to represent the sequences

$$h(n) = e^{-0.1n}u(n),$$
$$x(n) = (0.8)[u(n) - u(n - 5)].$$

Compile and test each routine. Using the routines developed, compute and print the convolution of *x(n)* and *h(n)* for $n = 0, \ldots, 99$.

The graphical interpretation of the convolution relationship given by Eqs. (3.15) and (3.16) is important. Consider the first form given by (3.15). For a particular value of the index *n*, say $n = n_1$, the output sample $y(n_1)$ can be computed by evaluating

$$y(n_1) = \sum_{k=-\infty}^{\infty} x(k)h(n_1 - k).$$

The actions necessary for computing the output sample $y(n_1)$ can be listed by inspection of the convolution sum as follows:

1. Sketch $x(k)$ as a function of k.
2. Sketch $h(n_1 - k)$ as a function of k.
3. Align the two sequences $x(k)$ and $h(n_1 - k)$ on the k-axis, and multiply the amplitudes of the overlapping samples.
4. Sum the products obtained in step 3. This yields the amplitude of one output sample at $n = n_1$.

Step 2 requires sketching $h(n_1 - k)$ as a function of k. This can be further broken down into several steps.

2a. Sketch $h(k)$ as a function of k.
2b. Fold (or flip) the sequence in step 2a around the origin to obtain $h(-k)$ as a function of k.

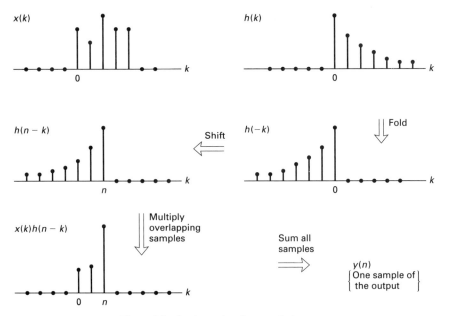

Figure 3.2 Implementing the convolution sum.

2c. Since $h(n_1 - k) = h(-[k - n_1])$, shift the sequence in step 2b to the right by n_1 samples to obtain $h(n_1 - k)$. Note that if n_1 is negative the shift is to the left.

Figure 3.2 graphically illustrates these steps.

Exercise 3.2.2

Consider the following two sequences:

$$x(n) = \{1.0, 2.0, 3.0, 4.0, 5.0\},$$
$$\uparrow$$
$$h(n) = \{1.0, 1.0, 1.0\}.$$
$$\uparrow$$

(a) Assuming that $h(n)$ is the impulse response of a DTLTI system, and $x(n)$ is the input signal applied to this system, determine the output signal $y(n)$ using the steps just outlined. Tabulate the nonzero range of $y(n)$. What is the length of the convolution result?

(b) Compute the output signal using PC-DSP. Convolution of the two sequences is accomplished by using the menu selections *Operations/Processing-functions/Convolve-sequences*. Tabulate the output sequence and compare with your hand calculations in part (a).

Exercise 3.2.3

Consider a linear time-invariant system with the impulse response $h(n) = e^{-0.1n}u(n)$. In this exercise, we will find the output $y(n)$ of this system when the input signal is in the form $x(n) = (0.8)[u(n) - u(n - 5)]$.

Sec. 3.2 Signal-system Interaction

(a) First, determine $y(0)$ using the first form of the convolution sum given in (3.15). On paper, sketch $x(k)$ and $h(0 - k)$ as functions of k. Compute $y(n)$ by multiplying the overlapping samples (only one in this case) and adding the products.

(b) Repeat the procedure in part (a) to compute $y(2)$ and $y(7)$. You will need to sketch the sequences $h(2 - k)$ and $h(7 - k)$.

(c) Repeat parts (a) and (b), this time using the alternative form of the convolution sum given by (3.16). Compute $y(0)$, $y(2)$, and $y(7)$ working with the sequences $x(0 - k)$, $x(2 - k)$, and $x(7 - k)$. How do the results compare to those found previously?

(d) Compute the output signal using PC-DSP. Even though the given impulse response $h(n)$ is of infinite length, using 100 samples for $n = 0, \ldots, 99$ should be sufficient. Tabulate and plot the output sequence. Check the output samples for $n = 0, 2,$ and 7.

Exercise 3.2.4

Consider the convolution problem in Exercise 3.2.3. It is also possible to find an analytical expression for the output signal $y(n)$. We will use the first form of the convolution sum given by (3.15). The expression for $y(n)$ should be determined for three different ranges of the index n. In each case, sketch the signals involved to verify the following statements:

(a) If $n < 0$, there is no overlap between $x(k)$ and $h(n - k)$, and therefore the output amplitude is equal to zero.

(b) If n is in the range $0 \leq n \leq 5$, then the range of overlapped samples starts at $k = 0$ and ends at $k = n$. Within this range of k, $x(k) = 0.8$, and $h(n - k) = e^{-0.1(n - k)}$. Thus, for values of n that satisfy $0 \leq n \leq 5$, the nth output sample should be computed by evaluating the convolution summation in the given overlap range of k; that is,

$$y(n) = \sum_{k=0}^{n} (0.8)e^{-0.1(n - k)}.$$

Using the techniques demonstrated in Section 2.2, a closed-form expression for $y(n)$ can be found. Show that

$$y(n) = 7.6067(1.1052 - e^{-0.1n}).$$

(c) If $n > 5$, then the range of overlapped samples starts at $k = 0$ and ends at $k = 5$. In this case, the nth output sample should be computed as

$$y(n) = \sum_{k=0}^{4} (0.8)e^{-0.1(n - k)},$$

which can be put into the closed form (verify)

$$y(n) = 4.9346e^{-0.1n}.$$

(d) Evaluate the analytical solution for the specific index values $n = 0, 2,$ and 7. Compare the results to corresponding results obtained in Exercise 3.2.3.

It is also possible to interpret the convolution equation in a slightly different perspective. An alternative approach based on the idea of superposition will be presented next, which should provide further insight into the process. Equation (3.15) is repeated here for convenience:

$$y(n) = \sum_{k=-\infty}^{\infty} x(k)h(n - k).$$

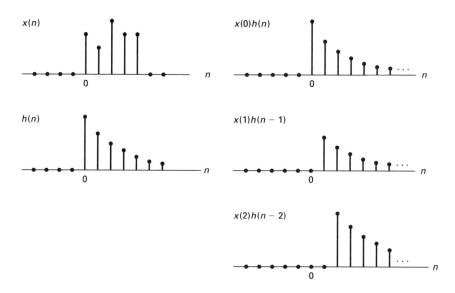

Figure 3.3 Implementing the convolution sum using superposition.

1. Sketch the impulse response $h(n)$ as a function of n.
2. For a particular value of k, shift the impulse response to the right by k samples to obtain $h(n - k)$. Note that if k is negative the actual shift is to the left.
3. Multiply all samples of $h(n - k)$ by the amplitude of the sample $x(k)$. The result of this is the sequence $x(k)h(n - k)$, sketched as a function of n. This is a shifted and scaled version of the impulse response. Recall that k is an integer constant.
4. Repeat steps 2 and 3 for all integer values of k.
5. Add all shifted and scaled impulse responses to obtain $y(n)$.

With this approach, the response of the system to each sample of the input sequence is considered individually, and the output $y(n)$ is obtained by adding the responses together. This amounts to expressing the input sequence $x(n)$ as a linear combination of shifted and scaled impulse sequences and considering the response of the system to each signal in the decomposition. The superposition interpretation of convolution is illustrated in Fig. 3.3.

Exercise 3.2.5

Repeat the requirements of Exercise 3.2.2, this time using the superposition approach just presented. Specifically, express $x(n)$ as the sum of five impulse sequences, and determine the response of the system to each. Find $y(n)$ by adding responses to components of $x(n)$.

Exercise 3.2.6

Consider again the system example used in Exercises 3.2.3 and 3.2.4. In this exercise, we will compute the output signal using the superposition approach. The input signal $x(n)$ can be expressed as a linear combination of scaled and shifted unit-impulse sequences; that is,

$$x(n) = 0.8\delta(n) + 0.8\delta(n - 1) + 0.8\delta(n - 2) + 0.8\delta(n - 3) + 0.8\delta(n - 4).$$

Sec. 3.2 Signal-system Interaction

The output sequence can be constructed by considering the response of the system to each component of $x(n)$ and adding these responses:

$$y(n) = 0.8h(n) + 0.8h(n - 1) + 0.8h(n - 2) + 0.8h(n - 3) + 0.8h(n - 4).$$

(a) Using PC-DSP, obtain the sequences $h(n)$, $h(n - 1)$, $h(n - 2)$, $h(n - 3)$, and $h(n - 4)$.

(b) Scale each sequence obtained in part (a) by 0.8, and then add the resulting sequences (two at a time) to obtain $y(n)$. Compare the resulting sequence $y(n)$ to that found in Exercise 3.2.3.

Exercise 3.2.7

In this exercise, we will focus our attention on determining the range of nonzero samples in the convolution of two finite-length sequences. Let $x(n)$ and $h(n)$ be two sequences that are equal to zero outside their respective index ranges; that is,

$$x(n) = 0, \quad \text{if } n < N_1 \text{ or } n > N_2,$$
$$h(n) = 0, \quad \text{if } n < N_3 \text{ or } n > N_4.$$

The convolution of these sequences is given by (3.15), which will be repeated here:

$$y(n) = \sum_{k=-\infty}^{\infty} x(k)h(n - k).$$

Even though the summation limits are infinite, the sequence $x(k)$ is equal to zero outside the range $k \in [N_1, N_2]$. Changing the summation limits to N_1 and N_2, respectively, will not affect the result. Therefore,

$$y(n) = \sum_{k=-N_1}^{N_2} x(k)h(n - k).$$

The term $h(n - k)$ is equal to zero outside the range $(n - k) \in [N_3, N_4]$. Complete the missing steps and show that

$$y(n) = 0, \quad \text{if } n < N_1 + N_3 \text{ or } n > N_2 + N_4.$$

On paper, determine the nonzero range of the convolution result for the sequence pairs given next. Afterward, use PC-DSP to generate and convolve these sequences. Use the results to check the accuracy of your hand calculations.

(a) $x(n) = u(n) - u(n - 10)$, and $h(n) = e^{-0.1n}[u(n - 2) - u(n - 8)]$.

(b) $x(n) = r(n - 1) - r(n - 7)$, and $h(n) = \sin(0.1n)[u(n + 5) - u(n - 15)]$.

Note: $r(n)$ represents the unit-ramp sequence defined as $r(n) = n\, u(n)$.

Exercise 3.2.8

Consider the following parallel combination of two linear and time-invariant systems. An equivalent system with impulse response $h_{eq}(n)$ can be found such that the input–output relationship remains unchanged.

(a) Working with the convolution sum, first obtain expressions for the signals $y_1(n)$ and $y_2(n)$ in terms of $x(n)$. After that, express the output signal $y(n)$ in terms of the input $x(n)$. Specifically, show that

$$h_{eq}(n) = h_1(n) + h_2(n).$$

Note that computer use is not necessary for this part of the problem.

(b) Consider a particular case where the two parallel systems have the impulse responses

$$h_1(n) = (0.8)^n u(n), \quad n = 0, \ldots, 19,$$
$$h_2(n) = 0.1[u(n) - u(n-10)].$$

Use PC-DSP to generate the impulse responses $h_1(n)$ and $h_2(n)$. Also obtain the impulse response $h_{eq}(n)$ of the equivalent system.

(c) Generate an arbitrary 100-sample input signal $x(n)$. Using the convolution function of PC-DSP, check the validity of the result in part (a). Compute the output $y(n)$ in two ways: (1) by computing $y_1(n)$ and $y_2(n)$ and adding them, and (2) by using the equivalent system. Compare the results.

Exercise 3.2.9

In this exercise, we will consider a series combination of two systems as opposed to a parallel combination. Again, the goal is to find an equivalent system.

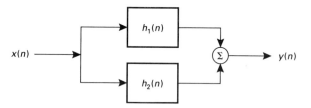

(a) Use the convolution sum to express the intermediate signal $w(n)$ in terms of the input $x(n)$. Afterward, express $y(n)$ in terms of $w(n)$ and show that

$$h_{eq}(n) = h_1(n) * h_2(n).$$

(b) Numerically verify this theoretical result using PC-DSP. Use the impulse responses given in Exercise 3.2.8. Use an arbitrary 100-sample input signal. Obtain the output signals of both the serial system and the equivalent system and compare.

(c) Reverse the order of connection for the two systems. Repeat part (b) of the exercise with the new configuration. How does the output signal compare to the previously obtained result?

Exercise 3.2.10

Consider again the series connection of two systems given in Exercise 3.2.9. The second block with impulse response $h_2(n)$ is to be replaced with a system that has the input–output relationship

$$y(n) = 0.1[w(n)]^2.$$

(a) Using the same input signal $x(n)$ that was used in Exercise 3.2.9, determine the output signal $y(n)$. Note that convolution techniques cannot be used to obtain $y(n)$ from $w(n)$. (Remember the constraints under which the convolution equation was derived.)

(b) Reverse the order of two blocks in the serial configuration, and compute the output signal. How does this affect the output of the system?

Sec. 3.3 System Function Concept   43

3.3 SYSTEM FUNCTION CONCEPT

In the previous section, we established the fact that a linear and time-invariant system can be completely described by its impulse response, since the impulse response is all that is needed to determine the system's output for any arbitrary input signal. Nevertheless, determining the output signal of a system in response to a particular input signal is not the only type of analysis problem of interest. In some cases, it is necessary to analyze the general behavior characteristics of a system without specific attention to a particular input signal, and alternative forms of describing a system are also used.

Let the input to a linear and time-invariant system be a complex exponential signal; that is,

$$x(n) = e^{j\omega_1 n}.$$

Using (3.16), the output of the system is

$$y(n) = \sum_{k=-\infty}^{\infty} h(k) e^{j\omega_1(n-k)}$$

which, after simple manipulation, can be written as

$$y(n) = e^{j\omega_1 n} \sum_{k=-\infty}^{\infty} h(k) e^{-j\omega_1 k}$$

$$= x(n) \sum_{k=-\infty}^{\infty} h(k) e^{-j\omega_1 k}.$$

This is a very important result because it states that the response of a linear time-invariant system to a complex exponential input signal is equal to the input sequence scaled by a complex number. Furthermore, this complex number depends on the impulse response of the system and the angular frequency of the complex exponential input sequence $x(n)$. Let $H(\omega)$ be defined as

$$H(\omega) = \sum_{k=-\infty}^{\infty} h(k) e^{-j\omega k}. \tag{3.17}$$

Thus, the response of the system to the complex exponential input signal is

$$y(n) = e^{j\omega_1 n} H(\omega_1).$$

$H(\omega)$ is a complex function of the angular frequency variable ω. It can be written in Cartesian or polar complex number forms:

$$H(\omega) = H_r(\omega) + jH_i(\omega) \tag{3.18}$$

or

$$H(\omega) = |H(\omega)| e^{j \arg[H(\omega)]}. \tag{3.19}$$

The equations for converting from one complex form into the other are

$$|H(\omega)| = \sqrt{H_r^2(\omega) + H_i^2(\omega)} \tag{3.20a}$$

$$\arg[H(\omega)] = \tan^{-1}\left[\frac{H_i(\omega)}{H_r(\omega)}\right], \quad (3.20b)$$

$$H_r(\omega) = |H(\omega)|\cos(\arg[H(\omega)]), \quad (3.20c)$$

$$H_i(\omega) = |H(\omega)|\sin(\arg[H(\omega)]). \quad (3.20d)$$

The complex function $H(\omega)$ is referred to as the *system function*. It is important to note that $H(\omega)$ as defined by (3.17) is a periodic function of ω with a period of 2π radians.

Exercise 3.3.1

Using the definition of the system function as given by (3.17), find system functions for the linear and time-invariant systems with the following impulse responses. Roughly sketch the magnitude and the phase of each system function.

(a) $h(n) = (0.95)^n u(n)$
(b) $h(n) = n(0.95)^n u(n)$
(c) $h(n) = (0.95)^n \cos(0.05\pi n) u(n)$
(d) $h(n) = u(n) - u(n - 10)$

Hints: For part (b), differentiate the summation obtained in part (a) and rearrange to obtain the desired result. For part (c), express the cosine function as a sum of two complex exponentials using Euler's formula.

Exercise 3.3.2

PC-DSP has a built-in function for computing the system function $H(\omega)$ of a system from the knowledge of the impulse response $h(n)$. This function can be accessed through the menu selections *Transforms/DTFT*. [DTFT stands for *discrete-time Fourier transform*, which will be the topic of the next chapter. For the moment, it will suffice to say that it computes the system function when applied to $h(n)$.] Using PC-DSP, compute the system functions for the systems described in Exercise 3.3.1. Plot the magnitude and the phase of each system function. Compare the results to your hand sketches.

Using the definition for $H(\omega)$, the output of the system in response to the complex exponential $e^{j\omega_1 n}$ is

$$y(n) = |H(\omega_1)| e^{j\arg[H(\omega_1)]} e^{j\omega_1 n}$$
$$= |H(\omega_1)| e^{j\{\omega_1 n + \arg[H(\omega_1)]\}}.$$

For a complex exponential input signal, the output of the system is also a complex exponential. It is obvious that, for a linear and time-invariant system, the complex function $H(\omega)$ characterizes the system's behavior when the input is a complex exponential. The system transforms the input complex exponential into the output complex exponential by scaling its magnitude and changing its phase. The magnitude scale factor is equal to the magnitude of the system function at the angular frequency of interest. Similarly, the added phase term is equal to the phase of the system function at the same angular frequency. Thus, $H(\omega)$ provides information about the frequency selective nature of the system, that is, how the system responds to input signals at different frequencies. The significance of this will be evident in Chapter 4 when decomposition of discrete-time signals into single-frequency components is discussed.

Sec. 3.3 System Function Concept

Exercise 3.3.3

The purpose of this exercise is to demonstrate the effects of a discrete-time linear and time-invariant system on a complex exponential input signal. The system that was analyzed in part (d) of Exercises 3.3.1 and 3.3.2 will be used. The impulse response of the system is

$$h(n) = u(n) - u(n - 10).$$

(a) Use PC-DSP to generate 100 samples of a complex exponential with unit magnitude and angular frequency $\omega_1 = 0.1\pi$, and convolve it with the system impulse response to obtain the output signal.

(b) Tabulate the output sequence. Pay attention to the steady-state region of the output sequence, that is, samples for $n \geq 10$. (Since the impulse response of the system is 10 samples long, "end effects" of convolution will only be cleared after 10 samples. Another way to see this is to realize that samples for $n < 10$ are computed using only a portion of the impulse response.)

(c) Within the steady-state region, pick an arbitrary sample of the output signal, say $y(45)$. Compute the magnitude and the phase of this sample. (Do this part by hand.) Repeat the same for the corresponding input sample, $x(45)$. How do the scaling and the additional phase relate to $|H(\omega_1)|$ and $\arg[H(\omega_1)]$ obtained in Exercise 3.3.2?

(d) Repeat parts (a) through (c) for angular frequencies $\omega_2 = 0.4\pi$ and $\omega_3 = 0.7\pi$.

It is sometimes of interest to find the response of a linear time-invariant system to a real-valued sinusoidal sequence, for example,

$$x(n) = \cos(\omega_1 n),$$

which, using Euler's formula, can be written as the sum of two complex exponential signals:

$$x(n) = 0.5e^{j\omega_1 n} + 0.5e^{-j\omega_1 n}.$$

Since the system under consideration is linear, its response to the sum of these terms is equal to the sum of responses to each term:

$$y(n) = 0.5H(\omega_1)e^{j\omega_1 n} + 0.5H(-\omega_1)e^{-j\omega_1 n}.$$

It can be shown (see Section 4.3) that, for a real-valued impulse response $h(n)$, the magnitude of the system function is an even function of ω,

$$|H(-\omega)| = |H(\omega)|$$

and the phase of the system function is odd,

$$\arg[-H(\omega)] = -\arg[H(\omega)].$$

Using these two properties, the response of the system to $x(n)$ is shown to be

$$y(n) = |H(\omega_1)|\cos(\omega_1 n + \arg[H(\omega_1)]).$$

Exercise 3.3.4

In this exercise, the we will repeat the steps of Exercise 3.3.3, but this time using a real-valued cosine sequence. Use the system with the impulse response

$$h(n) = u(n) - u(n - 10).$$

(a) Use PC-DSP to generate 100 samples of a cosine sequence with the angular frequency $\omega_1 = 0.1\pi$, that is,

$$x(n) = \cos(0.1\pi n)$$

and convolve it with the system impulse response to obtain the output signal.

(b) Plot the input and the output sequences on the same coordinate system using the *Graphics* menu selection. (You might want to use the continuous-plot option since viewing two discrete plots superimposed would be difficult.) Relate the amplitude and phase changes of the sinusoidal sequence to the magnitude and phase of the system function at ω_1.

(c) Repeat parts (a) and (b) for angular frequencies $\omega_2 = 0.4\pi$ and $\omega_3 = 0.7\pi$.

3.4 DIFFERENCE EQUATIONS

In the previous sections of this chapter, we considered two methods of describing a DTLTI system, the impulse response and the system function. We have established the fact that knowledge of the impulse response is sufficient for determining the response of the system to any arbitrary input signal. This is achieved by first decomposing the input signal into shifted and scaled impulses and then using the properties of linearity and time invariance. Alternatively, it will be shown in the next chapter that the knowledge of the frequency-domain system function is also sufficient to determine the response of the system to any arbitrary input signal.

A third method of description for a DTLTI system is a constant-coefficient difference equation, which can be written in the general form

$$\sum_{k=0}^{N} a_k y(n-k) = \sum_{r=0}^{M} b_r x(n-r), \qquad (3.21)$$

where a_k and b_r are constant coefficients. The difference equation describes the relationship between current and past (and perhaps future) samples of the input and the output sequences. It is possible to divide each term in (3.21) by the coefficient of the current output sample and rearrange the terms to obtain

$$y(n) = -\sum_{k=1}^{N} \bar{a}_k y(n-k) + \sum_{r=0}^{M} \bar{b}_r x(n-r), \qquad (3.22)$$

where

$$\bar{a}_k = \frac{a_k}{a_0}$$

and

$$\bar{b}_r = \frac{b_r}{a_0},$$

and it is assumed that $a_0 \neq 0$. (This is not a very restricting assumption. Even if $a_0 = 0$, the difference equation can be rewritten with a variable change applied to the index n so that the leading coefficient is nonzero.)

Sec. 3.4 Difference Equations

A DTLTI system can be uniquely described by specifying the coefficients of its difference equation. For a specified input signal $x(n)$, the latter form of the difference equation given by (3.22) also makes it possible to determine the output of the system one sample at a time. Consider the case $n = 0$. The difference equation can be written as

$$y(0) = -\sum_{k=1}^{N} \bar{a}_k y(-k) + \sum_{r=0}^{M} \bar{b}_r x(-r). \quad (3.23)$$

To compute $y(0)$, we need to know the past output samples that appear on the right side of (3.23); that is, we need

$$y(-1), y(-2), \ldots, y(-N).$$

These values are called the *initial conditions*. Based on knowledge of the input signal and the initial conditions, $y(0)$ can be determined. For the next output sample, we can write

$$y(1) = -\sum_{k=1}^{N} \bar{a}_k y(1-k) + \sum_{r=0}^{M} \bar{b}_r x(1-r). \quad (3.24)$$

The sample $y(0)$ now appears on the right side and is used for computing $y(1)$. This procedure can be repeated to compute other output samples $y(2), y(3), \ldots$, recursively.

In the implementation of discrete-time systems, sometimes the initial conditions may not be specified. In such cases, the system is assumed to be initially relaxed, and all initial conditions are assumed to be zero.

Exercise 3.4.1

Consider a DTLTI system described by the difference equation

$$y(n) = 1.79y(n-1) - 0.81y(n-2) + 0.89x(n-1).$$

(a) First, we will find the impulse response $h(n)$ of this system by iterating through the difference equation. This requires the discrete unit-impulse sequence be chosen as the input signal: $x(n) = \delta(n)$. Obviously, for a system with an infinite-length impulse response, this method cannot be used to determine $h(n)$ for all n, since each iteration produces one output sample.

We will assume that the system is causal so that no nonzero output samples can be seen before the only nonzero sample of the input signal appears at time $n = 0$. This also implies that the system is initially relaxed and all initial conditions are equal to zero.

Let's write the difference equation for $n = 0$:

$$y(0) = 1.79y(-1) - 0.81y(-2) + 0.89x(-1).$$

The initial conditions $y(-1)$ and $y(-2)$ were assumed to be zero. Additionally, since $x(n) = \delta(n)$, the sample $x(-1)$ is also equal to zero, and the output sample at $n = 0$ is $y(0) = 0$. Repeating this procedure for $n = 1$, we find

$$\begin{aligned} y(1) &= 1.79y(0) - 0.81y(-1) + 0.89x(0) \\ &= (1.79)(0) - (0.81)(0) + (0.89)(1) \\ &= 0.89. \end{aligned}$$

It is more convenient to continue the solution by building a table with entries for n, $y(n)$, $y(n-1)$, $y(n-2)$, and $x(n-1)$.

n	$y(n)$	$y(n-1)$	$y(n-2)$	$x(n-1)$
0	0	0	0	0
1	0.89	0	0	1
2	1.5931	0.89	0	0
3	1.4179	1.5931	0.89	0

Continue iterating through the difference equation in this fashion to obtain the first 10 samples of the impulse response of the system for $n = 0, \ldots, 9$.

(b) Using the same procedure, find the first 10 samples of the response to the input signal

$$x(n) = \{-0.2, 0.7, 1.2, 1.45, 0.9, 0.34\}.$$
$$\uparrow$$

Exercise 3.4.2

PC-DSP has a built-in function for evaluating a difference equation recursively. It is accessed with the menu selections *Operations/Processing-functions/Difference-equation*. The left side of the difference equation is assumed to be equal to $y(n)$, and only the right side is entered into the appropriate field. Any valid numerical expression can be used. For example, the entry

$$1.79*y(n-1)-0.81*y(n-2)+0.89*x(n-1)$$

represents the difference equation used in the previous exercise. Any necessary initial conditions are entered into the appropriate data fields. The user also enters the name of an existing input sequence, as well as initial and final index values for which the $y(n)$ is to be evaluated. If there are any initial conditions that are needed and not specified, they are assumed to be zero during the solution process, and a warning message is printed.

(a) Using the difference equation evaluation capability of PC-DSP, find the impulse response $h(n)$ of the system in Exercise 3.4.1 for $0 \leq n \leq 99$ by evaluating the difference equation. Assume that the system is initially relaxed. (Just leaving the initial conditions field blank accomplishes this.) Tabulate and plot the result, and compare its first 10 samples to those obtained in the previous exercise.

(b) Using PC-DSP, generate the input signal $x(n)$ used in part (b) of the previous exercise. The easiest method of doing this is to use the keyboard entry method accessed with the menu selections *Sequences/Generate-sequence/Read-from-keyboard*. Also, specify the starting index as 2.

(c) Find the response of the system to the input signal $x(n)$ for $0 \leq n \leq 99$. Tabulate and plot the result, and compare its first 10 samples to those obtained in the previous exercise.

Exercise 3.4.3

In this exercise, we will find an alternative solution for the system used in the previous two exercises. It is interesting to note that a difference equation can be written in a number of different forms, and the solution found depends on which form of it is used. The theory behind this will be reviewed in Chapter 6 when we discuss the z-transform. Consider the difference equation used in Exercise 3.4.1, which will be repeated here:

$$y(n) = 1.79y(n-1) - 0.81y(n-2) + 0.89x(n-1).$$

By rearranging terms, we can write this difference equation as

Sec. 3.4 Difference Equations

$$y(n-2) = -\frac{1}{0.81}y(n) + \frac{1.79}{0.81}y(n-1) + \frac{0.89}{0.81}x(n-1).$$

Making a variable change $\bar{n} = n - 2$ and calculating the coefficients, we obtain

$$y(\bar{n}) = -1.235 y(\bar{n}+2) + 2.21 y(\bar{n}+1) + 1.1 x(\bar{n}+1).$$

This alternative form of the difference equation can be used for evaluating $y(\bar{n})$ starting at $\bar{n} = 0$ and moving in the negative direction, that is, for $\bar{n} = -1, -2, -3, \ldots$.

(a) Assuming that the initial values $y(11)$ and $y(12)$ are both equal to zero, find the impulse response of this system for $-9 \leq \bar{n} \leq 10$. Compare the result to that obtained in Exercise 3.4.1. Based on your comparison, does the alternative form of the difference equation still represent the same system as in Exercise 3.4.1, or does it correspond to a different system? This part of the exercise does not require the use of a computer.

(b) Use PC-DSP to find the impulse response in the range $-99 \leq \bar{n} \leq 100$. Note that care must be taken in specifying the range of the index for which the left side is to be computed. PC-DSP computes the left side starting with the initial index value and going up or down to the final index value. Specify these to be 100 and -99, respectively, and not the other way around. Otherwise, $y(-99)$ would need to be computed first, without the knowledge of $y(-98)$ and $y(-97)$. Also, the initial values $y(101)$ and $y(102)$ may be specified if desired. (Default is zero.) Tabulate and plot the resulting sequence.

It is obvious from the foregoing discussion that a difference equation can be used to implement a DTLTI system in real time. Recall that real-time processing means that the signal is processed as it occurs; that is, the system gets the input signal one sample at a time and must respond to each incoming sample with an output sample of its own. Various methods of implementation will be discussed further in Chapter 6. For real-time processing, the difference equation must be in the form given in Exercise 3.4.2 so that output samples can be computed for successive values of the index n in ascending order. If the difference equation is in the alternative form given in Exercise 3.4.3, then each output sample depends on future input and output samples.

Exercise 3.4.4

Consider the running integral of a continuous function:

$$y_a(t) = \int_{-\infty}^{t} x_a(t)\, dt. \quad (3.25)$$

At a given time instant t, the value of the integral is equal to the area under the function $x_a(t)$ up to that instant. A numerical approximation to this integral can be obtained by evaluating the integrand at regular intervals and assuming that the function value remains constant between successive evaluations. Thus, the area under the function is approximated using areas of rectangles formed using samples of the signal (rectangular approximation). Figure 3.4 depicts this situation.

Thus, the running integral in (3.25) is approximated with a running sum:

$$y_a(nT) \approx \sum_{k=-\infty}^{n} T x_a(kT), \quad (3.26)$$

where T is the time step used in the approximation. If we define two discrete-time sequences $x(n)$ and $y(n)$ as

$$x(n) = x_a(nT)$$

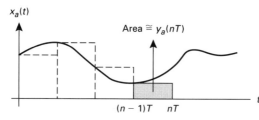

Figure 3.4 Rectangular approximation of a function.

and
$$y(n) = y_a(nT),$$
then the rectangular approximation to the running integral can be written as
$$y(n) \approx T \sum_{k=-\infty}^{n} x(k). \qquad (3.27)$$
Even though the running sum of (3.27) can be thought of as a difference equation, its right side has an infinite number of terms, and thus its practical usefulness as a means of computing an approximation to the integral of (3.25) is somewhat limited. If we evaluate $y(n-1)$ using (3.27), we obtain
$$y(n-1) \approx T \sum_{k=-\infty}^{n-1} x(k),$$
which can be substituted into (3.27) to yield
$$y(n) \approx y(n-1) + Tx(n). \qquad (3.28)$$

(a) We will use the technique outlined to approximate the integral of the function $x_a(t) = e^{-t}u(t)$. Use the time step $T = 0.1$ and generate samples of the sequence $x(n)$ for $n = 0, \ldots, 50$. (Note that this corresponds to the time interval $0 \leq t \leq 5$ s.) Then use the difference equation evaluation function of PC-DSP to evaluate (3.28). This results in a sequence of *approximate values* of the integral.

(b) Compute the actual integral in (3.25) analytically to obtain $y_a(t)$. Evaluate this result at time instants $t = nT$ to obtain *exact values* of the integral at those instants. Compute the approximation error by subtracting the sequence found in part (a) from the sequence of exact values. Tabulate and graph this approximation error.

(c) Repeat parts (a) and (b) using a time step of $T = 0.01$ s and evaluating approximate and exact values for $n = 0, \ldots, 500$. Tabulate and graph the approximation error and compare to that obtained previously.

Exercise 3.4.5

For the evaluation of a running integral, an alternative to the rectangular approximation given by (3.26) is the *trapezoidal approximation*,
$$y_a(nT) \approx \sum_{k=-\infty}^{n} \frac{T}{2}[x_a(kT) + x_a(kT - T)], \qquad (3.29)$$
which amounts to approximating the area under the function with the areas of the trapezoids formed between successive evaluations of the function. This is illustrated in Fig. 3.5.

(a) Using the techniques detailed in Exercise 3.4.6, show that a workable difference equation can be found in the form

Sec. 3.4 Difference Equations

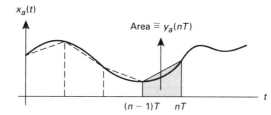

Figure 3.5 Trapezoidal approximation of a function.

$$y(n) \approx y(n-1) + \frac{T}{2}[x(n) + x(n-1)]. \tag{3.30}$$

(b) Repeat parts (a), (b), and (c) of Exercise 3.4.4 using trapezoidal approximation instead of rectangular approximation. How do the approximation errors compare? Is this a better approximation? Comment.

Exercise 3.4.6

The difference equations used in the preceding exercises are all linear constant-coefficient difference equations that represent DTLTI systems. It is possible to prove that the difference equation of a linear and time-invariant system must also be linear and time invariant, and therefore its right side must be a linear combination of input and output samples. In this exercise, we will consider an example of a nonlinear difference equation used in digital computers as well as hand-held calculators for finding the square root of a real positive number. The function

$$f(\alpha) = \alpha^2 - A$$

is equal to zero at the point $\alpha = \sqrt{A}$ if A is a positive number. Using Newton's method for finding real zeros of functions, we may start with an initial guess α_{old} and find an improved guess α_{new} by

$$\alpha_{new} = \alpha_{old} - \frac{f(\alpha_{old})}{f'(\alpha_{old})}. \tag{3.31}$$

Geometrically, this amounts to drawing a line that is tangent to the function $f(\alpha)$ at the point $\alpha = \alpha_{old}$ and using the intersection of the tangent line with the horizontal axis as the improved guess α_{new}. (See Fig. 3.6.) The solution can be obtained by repeated use of (3.31) until the difference between two successive values is insignificant. We will make the following definitions:

$$y(n) = \alpha_{new} \quad \text{and} \quad y(n-1) = \alpha_{old}.$$

Realizing that

$$f(\alpha_{old}) = y^2(n-1) - A$$

and

$$f'(\alpha_{old}) = 2y(n-1),$$

the following nonlinear difference equation can be written (verify):

$$y(n) = \frac{1}{2}\left[y(n-1) + \frac{A}{y(n-1)}\right]. \tag{3.32}$$

(a) Using the difference equation evaluation function of PC-DSP, solve this difference equation to find the square root of $A = 5$. Evaluate $y(n)$ for $n = 0, \ldots, 20$. Use the initial condition $y(-1) = 1$. Tabulate the resulting sequence. Does it progressively approach the correct value?

(b) Repeat part (a) starting with the initial value $y(-1) = 25$. How does this affect the convergence?

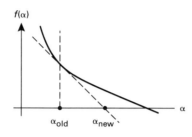

Figure 3.6 Newton's method for finding real zeros.

Exercise 3.4.7

(a) Using the development in the previous exercise as a model, write a difference equation for finding the cube root of a number.

(b) Using PC-DSP, solve the difference equation found in part (a) to find the cube root of $A = 125$. Use the initial condition $y(-1) = 10$. Evaluate $y(n)$ for $n = 0, \ldots, 20$.

(c) Repeat part (a) starting with the initial value $y(-1) = -100$. How does this affect the convergence?

Exercise 3.4.8

Similar to a running integral, the differential of a function can also be approximated using discrete-time signal-processing techniques. Let $y_a(t)$ be defined as

$$y_a(t) = \frac{dx_a(t)}{dt}.$$

At a given time instant t, the value of the time derivative of $x_a(t)$ is equal to the slope of the line that is tangent to $x_a(t)$ at that point. A well-known approximation is

$$y_a(nT) \approx \frac{x_a(nT) - x_a(nT - T)}{T},$$

which can be expressed with a difference equation:

$$y(n) \approx \frac{1}{T} [x(n) - x(n-1)]. \tag{3.33}$$

This is known as the *first backward difference* formula.

(a) Using the technique outlined, approximate the time derivative of the function $x_a(t) = \sin(t)u(t)$. Use the time step $T = 0.1$ and generate samples of the sequence $x(n)$ for $n = 0, \ldots, 50$ corresponding to the time interval $0 \le t \le 5$ s. Then use PC-DSP to evaluate (3.33). This results in a sequence of *approximate values* of the derivative.

(b) Compute the actual derivative of $x_a(t)$ analytically to obtain $y_a(t)$. Evaluate this result at time instants $t = nT$ to obtain *exact values* of the derivative at those instants. Compute the approximation error by subtracting the sequence found in part (a) from the sequence of exact values. Tabulate and graph this approximation error.

(c) Repeat parts (a) and (b) using a time step of $T = 0.01$ s and evaluating approximate and exact values for $n = 0, \ldots, 500$. Tabulate and graph the approximation error and compare to that obtained previously.

Exercise 3.4.9

A widely used method of smoothing a sequence is to process it through a system such that each output sample is the arithmetic average of the most recent M input samples. This can be thought of

Sec. 3.5 Stability and Causality

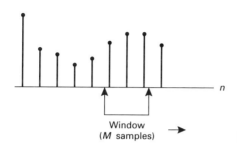

Figure 3.7 Moving-average filtering.

as looking at the sequence through a window that is M sample intervals wide, and averaging the samples that are visible. The window is moved to the right with each new input sample so that the most recent M samples are visible at any time. Because of this, a discrete-time system based on this idea is termed an *M-tap moving-average filter*. The situation is illustrated in Fig. 3.7. The difference equation for such a filter is

$$y(n) = \frac{1}{M} \sum_{k=0}^{M-1} x(n-k). \qquad (3.34)$$

(a) A sequence with the name *ex3-4-9* is on your distribution disks. First, graph this sequence. Process it using a three-tap moving-average filter and graph the output. Is the smoothing effect visible?

(b) Repeat part (a) with $M = 5$ and then with $M = 8$. What seems to be the correlation between the size of the moving-average window and the degree of smoothing? Comment.

(c) Can you think of a signal that would be completely blocked by a three-tap moving-average filter? *Hint*: Think of what happens when the three-sample window is moved at each step. One sample falls out from the left, and a new one enters from the right.

Exercise 3.4.10

Consider a linear time-invariant system described by the difference equation

$$y(n) = y(n - N) + x(n).$$

(a) Using PC-DSP, compute the first 150 samples of the impulse response of this system with $N = 10$. Assume that the system is initially relaxed. Tabulate and graph the result.

(b) Repeat part (a) with $N = 20$. How does this change affect the impulse response? Can this system be used for generating a periodic sequence from a finite-length input signal? How?

(c) Generate a 10-sample sequence

$$x(n) = (9 - n)u(n), \qquad n = 0, \ldots, 9.$$

Process it with the system described, using $N = 10$. Compute and graph the first 150 samples of the resulting output sequence.

(d) Repeat part (c) with $N = 20$.

3.5 STABILITY AND CAUSALITY

A system is said to be *stable* if, for every bounded input signal, the output signal is also bounded. Mathematically, the stability of a system implies that two finite positive numbers A and B can be found such that, if $|x(n)| \leq A$ for all integers n, then $|y(n)| \leq B$ for all integers n. In some textbooks, this definition of stability is referred to as the BIBO (bounded input bounded output) stability criterion, and other definitions are also given, especially in the area of control systems. In basic signal-processing applications, however, other stability definitions are usually not needed, and the BIBO definition will suffice.

If the system under consideration is linear and time invariant, then its output signal $y(n)$ can be expressed as the convolution of the input signal $x(n)$ and the impulse response $h(n)$. The absolute value of the output signal is

$$|y(n)| = \left| \sum_{k=-\infty}^{\infty} x(n-k)h(k) \right|.$$

The absolute value of a sum is less than or equal to the sum of absolute values of its terms, and the following inequality can be written:

$$|y(n)| \leq \sum_{k=-\infty}^{\infty} |x(n-k)| |h(k)|.$$

Replacing $|x(n-k)|$ by A, we get

$$|y(n)| \leq A \sum_{k=-\infty}^{\infty} |h(k)|.$$

Thus, for $|y(n)|$ to be bounded, we require

$$\sum_{k=-\infty}^{\infty} |h(k)| < \infty \qquad (3.35)$$

For a linear and time-invariant system to be stable, the impulse response must be absolutely summable.

From a practical implementation point of view, a discrete-time system to be designed and implemented is almost always required to be stable. Theoretically, an unstable system might produce output samples with infinite amplitude, even though the input signal might be well behaved. In practice, very large numbers are generated that cause floating-point overflow problems, since they cannot be represented within the binary numbering scheme of the processor.

Exercise 3.5.1

Impulse responses of a several discrete-time linear time-invariant systems are given next. Checking the stability condition given by (3.35) on each system, determine which systems are stable and which are not. Roughly sketch each impulse response. You might also want to plot a finite portion of each sequence using PC-DSP to see its general shape. To do this, use the menu selections *Sequences/Generate-sequence/Formula-entry-method*, and enter the analytical form of the impulse response. Samples generated in the range $-100 \leq n \leq 100$ should be sufficient to check your hand sketches.

(a) $h(n) = (0.8)^n$, all n
(b) $h(n) = (0.8)^n u(n)$
(c) $h(n) = (0.8)^n u(-n)$
(d) $h(n) = (1.2)^n u(-n)$
(e) $h(n) = \dfrac{(1.2)^n}{n} u(n)$

Sec. 3.5 Stability and Causality

Exercise 3.5.2

Several discrete-time systems are described next through their difference equations. Comment on the stability of each system. Afterward, use the difference equation evaluation function of PC-DSP to check your results. Specifically, find a portion of the impulse response of each system and check it in light of the stability condition in (3.35).

(a) $y(n) = y(n-1) + Tx(n)$ (rectangular integrator)

(b) $y(n) = y(n-1) + \frac{T}{2}[x(n) + x(n-1)]$ (trapezoidal integrator)

(c) $y(n) = 0.8y(n-1) + x(n)$

(d) $y(n) = x(n) + 3x(n-1) + 2x(n-2)$

Exercise 3.5.3

Consider a nonlinear discrete-time system described by the difference equation

$$y(n) = x(n)[y(n-1) + 1].$$

(a) Assume that the system is initially relaxed. Using the difference equation evaluation function of PC-DSP, find the first 500 samples of the impulse response of this system. Does the impulse response seem to satisfy (3.35)?

(b) This time use a unit-step sequence as input, and find the first 500 samples of the response of the system. Does the response look like one that might be expected from a stable system? If not, is there a discrepancy between the results of parts (a) and (b)? How would you explain it? *Hint*: Think about the assumptions made in deriving (3.35).

A system is said to be *causal* if, at any time, its output depends only on the values of present and past input samples and, possibly, past output samples. The current output sample of a causal system does not depend on any future input or output values. A system that does not satisfy causality requirements is called a *noncausal* system. In such a system, a knowledge of future events is necessary for computing the current output sample.

For obvious reasons, a system must be causal if it will be implemented in real time. This is an important consideration in system design problems. On the other hand, noncausal systems can be implemented in postprocessing mode. In this case, the entire signal might be recorded in advance. Thus, when determining the response of the system for a certain value of the index, future input samples can be available.

The output of a DTLTI system in response to an input signal $x(n)$ is determined by the convolution equation

$$y(n) = \sum_{k=-\infty}^{\infty} h(k)x(n-k).$$

For the system to be causal, future input samples (that is, input samples with index values greater than the current index n) should not appear on the right side. This, in turn, requires that the impulse response $h(k)$ be zero for negative values of k. Intuitively, this makes sense, since a causal system cannot start responding to an impulse sequence before the only nonzero sample of the impulse sequence is seen.

Exercise 3.5.4

(a) Consider a DTLTI system with the difference equation

$$y(n) = x(n+2) + 0.75x(n+1) + 0.5x(n) + 0.25x(n-1).$$

Assume that the system is initially at rest. Can this system be implemented in real time? Specifically, assume that, at sample index $n = 0$, the input sample $x(0) = 1.7$ is received.

(b) One method to make a noncausal system causal is to delay the input signal by an appropriate number of samples. Consider the following block diagram shown in which the noncausal system is cascaded with an M-sample delay unit.

What is the minimum delay necessary for the system under consideration? Write the difference equation between the input and the output signals of the modified system.

(c) Now consider a system with the difference equation

$$y(n) = 0.5y(n+1) - 0.5y(n-1) + x(n) + x(n-1).$$

Can this system be made causal by application of the technique described? Comment.

Exercise 3.5.5

The impulse response of a DTLTI system is

$$h(n) = \{0.2, 0.4, 0.6, 0.8, 1.0, 0.8, 0.6, 0.4, 0.2\}.$$
$$\quad\uparrow$$

(a) Compute the unit-step response of this system by generating the first 100 samples of a unit-step sequence and convolving it with $h(n)$. At what value of the sample index does the system start responding to the unit-step sequence?

(b) Find a causal system with impulse response $h_2(n) = h(n - M)$ and repeat part (a).

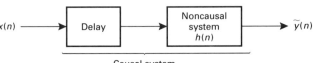

Can you compute the output sample $y(0)$? If the answer is no, then what is the earliest sample index at which $y(0)$ can be computed?

4

Frequency-domain Analysis

> Everything should be as simple as possible—
> but not simpler. —Albert Einstein

In Section 3.3, the system function concept was introduced as an alternative means of describing a linear and time-invariant system. The system function approach makes it possible to analyze a system in terms of its responses to sinusoidal signals at varying frequencies. The same idea can be put to use for analyzing signals in the frequency domain. Most discrete-time signals can be expressed as a linear combination of sinusoidal or complex exponential signals with various angular frequencies. In signal-system interaction problems involving DTLTI systems, decomposition of the input signal into sinusoidal or complex exponential components makes it possible to examine the effects of the system on each component. We have established in Chapter 3 that the response of a DTLTI system to a sinusoidal signal is another sinusoidal signal with modified amplitude and phase. If the input signal is decomposed into sinusoidal components and if the response of the system to each component is known, then the output signal can be constructed by modifying each component of the input signal as dictated by the system function. In this chapter, we will consider frequency-domain analysis methods for discrete-time signals as well as systems.

4.1 DISCRETE-TIME FOURIER TRANSFORM

The discrete-time Fourier transform (DTFT) of a sequence $x(n)$ is defined as

$$X(\omega) = \sum_{n=-\infty}^{\infty} x(n)e^{-j\omega n}. \tag{4.1}$$

Note that this definition is identical to the system function definition given in (3.17); thus, a system function is simply the DTFT of the impulse response of the system. As defined in (4.1), $X(\omega)$ is a continuous and generally complex function of the angular frequency variable ω. Due to the periodicity of the complex exponential term, $X(\omega)$ is a periodic function of ω with a period of 2π radians; that is,

$$X(\omega + 2k\pi) = X(\omega)$$

for any integer k. Because of this property, the DTFT of a sequence is usually specified over a 2π radian range of ω, and the rest of it can be deduced from this information. Like any complex function, $X(\omega)$ can be written in either Cartesian or polar complex number forms:

$$X(\omega) = X_r(\omega) + jX_i(\omega) \tag{4.2}$$

or

$$X(\omega) = |X(\omega)|e^{j\arg[X(\omega)]}. \tag{4.3}$$

Conversion from one complex form to the other is done through the following equations:

$$|X(\omega)| = \sqrt{X_r^2(\omega) + X_i^2(\omega)}, \tag{4.4a}$$

$$\arg[X(\omega)] = \tan^{-1}\left[\frac{X_i(\omega)}{X_r(\omega)}\right], \tag{4.4b}$$

$$X_r(\omega) = |X(\omega)|\cos(\arg[X(\omega)]), \tag{4.4c}$$

$$X_i(\omega) = |X(\omega)|\sin(\arg[X(\omega)]). \tag{4.4d}$$

The DTFT as defined in (4.1) is a summation with an infinite number of terms, and therefore, may or may not exist for a given signal. It can be shown that a necessary and sufficient condition for the DTFT of a signal $x(n)$ to exist is that $x(n)$ be absolutely summable; that is,

$$\sum_{n=-\infty}^{\infty} |x(n)| < \infty. \tag{4.5}$$

If the transform $X(\omega)$ is given, then $x(n)$ can be found using the *inverse-DTFT* equation

$$x(n) = \frac{1}{2\pi} \int_{-\pi}^{\pi} X(\omega)e^{j\omega n}d\omega, \tag{4.6}$$

where, in contrast with the continuous-time inverse Fourier transform, the integral is evaluated in the range $(-\pi, \pi)$. This is consistent with the fact that the DTFT of a signal is unique only within a 2π radian interval. Together, Eqs. (4.1) and (4.6) form a transform pair.

Sec. 4.1 Discrete-time Fourier Transform

TABLE 4.1 Some Commonly Used DTFT Pairs

Signal		Transform
1. $\delta(n)$	$\Longleftrightarrow$	$1, \quad -\infty < \omega < \infty$
2. $\delta(n - M)$	$\Longleftrightarrow$	$e^{-j\omega M}$
3. $u(n)$	$\Longleftrightarrow$	$\dfrac{1}{1 - e^{-j\omega}} + \pi \sum\limits_{k=-\infty}^{\infty} \delta(\omega - 2\pi k)$
4. $a^n u(n), \; \|a\| < 1$	$\Longleftrightarrow$	$\dfrac{1}{1 - ae^{-j\omega}}$
5. $e^{j\omega_0 n}$	$\Longleftrightarrow$	$2\pi \sum\limits_{k=-\infty}^{\infty} \delta(\omega - \omega_0 + 2\pi k)$
6. $\sin(\omega_0 n)$	$\Longleftrightarrow$	$\dfrac{\pi}{j} \sum\limits_{k=-\infty}^{\infty} [\delta(\omega - \omega_0 + 2\pi k) + \delta(\omega - \omega_0 + 2\pi k)]$
7. $\cos(\omega_0 n)$	$\Longleftrightarrow$	$\pi \sum\limits_{k=-\infty}^{\infty} [\delta(\omega - \omega_0 + 2\pi k) + \delta(\omega - \omega_0 + 2\pi k)]$
8. $u(n) - u(n - M)$	$\Longleftrightarrow$	$\dfrac{\sin(\omega M / 2)}{\sin(\omega / 2)} e^{-j(M-1)/2}$

Table 4.1 lists some commonly used DTFT pairs. Some of the entries in this table can be easily derived by direct application of the DTFT definition in (4.1). A few entries are easier to derive after going through the next section, where properties of the DTFT are given. Also, note that the discrete-time Fourier transforms for some signals that do not satisfy the existence condition in (4.5) are listed in this table. (See entries 3, 5, 6, and 7.) This is only possible if we are willing to resort to the use of continuous impulse functions within the transform.

Exercise 4.1.1

(a) Analytically determine the DTFT of the signal

$$x(n) = a^n u(n), \qquad |a| < 1$$

using the definition of DTFT along with the geometric series closed-form formula presented in Section 2.2. Check your result against entry 4 of Table 4.1. How does the convergence criterion for the geometric series relate to the existence condition for the DTFT?

(b) Write the expressions for the magnitude and the phase of the transform. For $a = 0.8$, roughly sketch the magnitude and the phase as functions of ω.

(c) Using PC-DSP, generate the signal $x(n)$ for the range of the index $n = 0, \ldots, 99$. [*Note:* This is only an approximation to the signal $x(n)$; however, since sample amplitudes of $x(n)$ decay rapidly, the samples we omit should be negligible.]

(d) Using the DTFT function of PC-DSP accessed with menu selections *Transforms/DTFT*, compute the DTFT of the sequence obtained in part (c). Note that PC-DSP computes the transform for $-\pi \le \omega \le \pi$. Once computed, the transform can be graphed by pressing the F2 key. It can also be tabulated at angular frequency increments of $\pi/256$ radians using the F3 key. Compare the results to your hand sketches.

Exercise 4.1.2

(a) On paper, verify entry 8 of Table 4.1; that is, find an expression for the DTFT of the signal

$$x(n) = u(n) - u(n - M).$$

Hint: Start with the DTFT definition. Realizing that $x(n)$ has a finite number of nonzero samples, write the transform as a finite summation with lower limit equal to zero and upper limit equal to $M - 1$. Factor out proper exponential terms from the numerator and the denominator such that the remaining exponentials can be written as sine functions using Euler's formula.

(b) Write the expressions for the magnitude and the phase of the transform. Note that the transform in Table 4.1 is *almost* in polar complex form with "magnitude"

$$\frac{\sin(\omega M/2)}{\sin(\omega/2)}$$

and phase

$$-\omega\left(\frac{M-1}{2}\right).$$

The only problem is that the ratio of two sinusoids could be negative for certain values of ω, so the first expression is not quite the magnitude. This can be resolved by changing negative values in the "magnitude" function to positive values and accounting for the lost minus signs in the phase function (by adding π to the phase at frequencies where the sign change is made). Keeping this in mind, roughly sketch the magnitude and the phase of the transform for $M = 3, 5,$ and 8.

(c) Using PC-DSP, generate the signal $x(n)$ for $M = 3$ and then compute the transform. Graph its magnitude and phase and compare to your pencil and paper solution.

(d) Repeat part (c) for $M = 5$ and $M = 8$.

(e) Recall that a scaled version of $x(n)$ was used in Exercise 3.4.8 as the impulse response of an M-tap averaging filter, and its application to a signal resulted in a smoothing effect the degree of which was dependent on M. How would you explain that smoothing effect after observing the magnitude of the transform for different values of M?

4.2 PROPERTIES OF THE DISCRETE-TIME FOURIER TRANSFORM

The DTFT has a number of interesting properties that are useful for simplifying derivations and for providing further insight into the concept of the frequency-domain analysis of discrete-time signals and systems. In this section, important properties of the DTFT will be summarized. To keep the notation simple, we will denote the forward transform with

$$X(\omega) = \text{DTFT}\{x(n)\}$$

and the inverse transform with

$$x(n) = \text{DTFT}^{-1}\{X(\omega)\}.$$

The notation

$$x(n) \iff X(\omega)$$

will be employed to indicate that $x(n)$ and $X(\omega)$ are a transform pair. Proofs of most of these properties will be left to the reader. All properties can be easily proved starting with the definition of the DTFT given by (4.1). Table 4.2 lists some of the important DTFT properties.

Sec. 4.2 Properties of the Discrete-time Fourier Transform

TABLE 4.2 Some Properties of the DTFT

Signal		Transform
1. $x(n)$	$\Longleftrightarrow$	$X(\omega)$
2. $x^*(n)$	$\Longleftrightarrow$	$X^*(-\omega)$
3. Linearity		
$\alpha x_1(n) + \beta x_2(n)$	$\Longleftrightarrow$	$\alpha X_1(\omega) + \beta X_2(\omega)$, all α and β
4. Time shifting		
$x(n - M)$	$\Longleftrightarrow$	$e^{-j\omega M} X(\omega)$
5. Time reversal		
$x(-n)$	$\Longleftrightarrow$	$X(-\omega)$
6. Frequency shifting		
$e^{j\omega_0 n} x(n)$	$\Longleftrightarrow$	$X(\omega - \omega_0)$
7. Modulation theorem		
$x(n)\cos(\omega_0 n)$	$\Longleftrightarrow$	$\frac{1}{2} X(\omega - \omega_0) + \frac{1}{2} X(\omega + \omega_0)$
$x(n)\sin(\omega_0 n)$	$\Longleftrightarrow$	$\frac{1}{2j} X(\omega - \omega_0) - \frac{1}{2j} X(\omega + \omega_0)$
8. Convolution		
$x(n)*h(n)$	$\Longleftrightarrow$	$X(\omega)H(\omega)$
9. Differentiation in frequency		
$nx(n)$	$\Longleftrightarrow$	$j\dfrac{dX(\omega)}{d\omega}$
10. Conjugate symmetric sequence		
$x(n) = x^*(-n)$	$\Longleftrightarrow$	$X(\omega)$ real
11. Conjugate antisymmetric sequence		
$x(n) = -x^*(-n)$	$\Longleftrightarrow$	$X(\omega)$ imaginary
12. $x(n)$ real	$\Longleftrightarrow$	$\|X(\omega)\| = \|X(-\omega)\|$
		$\arg[X(\omega)] = -\arg[X(-\omega)]$

Linearity

The DTFT is a linear transform. Given two DTFT pairs

$$x_1(n) \Longleftrightarrow X_1(\omega)$$

and

$$x_2(n) \Longleftrightarrow X_2(\omega)$$

and two constants α and β, it can be shown that

$$\alpha x_1(n) + \beta x_2(n) \Longleftrightarrow \alpha X_1(\omega) + \beta X_2(\omega). \tag{4.7}$$

Exercise 4.2.1

In this exercise, we will numerically verify the linearity property of the DTFT using the signals generated in the previous two exercises. We will refer to the signal generated in Exercise 4.1.1 as $x_1(n)$, and the signal generated in Exercise 4.1.2 for $M = 3$ will be called $x_2(n)$.

(a) Use PC-DSP to obtain the sum of these two signals: $x(n) = x_1(n) + x_2(n)$.
(b) Compute the DTFT of $x(n)$. Tabulate it along with the individual transforms of $x_1(n)$ and $x_2(n)$.
(c) Pick a few arbitrary values for ω and show for each that the transform of the sum of two sequences is equal to the sum of two transforms; that is,

$$X(\omega) = X_1(\omega) + X_2(\omega).$$

Time Shifting

For an arbitrary DTFT pair

$$x(n) \Longleftrightarrow X(\omega),$$

shifting the signal $x(n)$ in the time domain is equivalent to multiplying the transform by a complex exponential; that is,

$$x(n - M) \Longleftrightarrow e^{-j\omega M} X(\omega) \qquad (4.8)$$

The significance of (4.8) is that shifting (delaying) the signal does not affect the magnitude of the frequency spectrum. It simply increases the phase by ωM; resulting in

$$|\text{DTFT}\{x(n - M)\}| = |\text{DTFT}\{x(n)\}|$$

and

$$\arg[\text{DTFT}\{x(n - M)\}] = \arg[\text{DTFT}\{x(n)\}] - \omega M$$

Exercise 4.2.2

Consider the signal $x(n)$ generated in part (c) of Exercise 4.1.2, which will be repeated here for convenience:

$$x(n) = u(n) - u(n - 3)$$

(a) A new signal $w(n)$ is constructed by shifting $x(n)$ to the right by two samples:

$$\begin{aligned} w(n) &= x(n - 2) \\ &= u(n - 2) - u(n - 5) \end{aligned}$$

Generate the signal $w(n)$ using the *Shift-sequence* function of PC-DSP. This function is accessed through menu selections *Operations/Arithmetic-operations/Shift*. Do not forget to specify *linear* shift.

(b) Compute the DTFT of this new signal $w(n)$. Graph its magnitude and phase. Is the magnitude of the transform affected by the delay? Is the phase affected?

(c) Tabulate the transforms of $x(n)$ and $w(n)$ simultaneously. Pick a few arbitrary values for ω and numerically verify the time-shifting property of the DTFT.

Time Reversal

Given a DTFT pair

$$x(n) \Longleftrightarrow X(\omega),$$

time reversing the sequence $x(n)$ causes its transform to be reversed in the frequency domain; that is,

$$x(-n) \Longleftrightarrow X(-\omega). \qquad (4.9)$$

Exercise 4.2.3

Consider again the signal $x(n)$ generated in part (c) of Exercise 4.1.2. Its analytical form was

$$x(n) = u(n) - u(n - 3).$$

Sec. 4.2   Properties of the Discrete-time Fourier Transform

(a) A new signal $w(n)$ is defined as the time-reversed version of $x(n)$:
$$w(n) = x(-n)$$
$$= u(-n) - u(-n-3).$$

Generate this signal using the *Flip-sequence* function of PC-DSP accessed using the menu selections *Operations/Arithmetic-operations/Flip-sequence*.

(b) Compute the DTFT of this new signal $w(n)$. Graph and observe the magnitude and the phase of the transform. Are they consistent with Eq. (4.9)?

(c) Tabulate the transforms of $x(n)$ and $w(n)$ simultaneously. Pick a few arbitrary values for ω and numerically verify the time reversal property of the DTFT.

Frequency Shifting

For an arbitrary DTFT pair
$$x(n) \Longleftrightarrow X(\omega),$$
it can be shown that multiplying the signal $x(n)$ by a complex exponential sequence in the time domain is equivalent to shifting the transform to the right by an amount equal to the angular frequency of that complex exponential sequence; that is,
$$e^{j\omega_0 n} x(n) \Longleftrightarrow X(\omega - \omega_0). \tag{4.10}$$

Modulation Theorem

Given a DTFT pair
$$x(n) \Longleftrightarrow X(\omega),$$
it can be shown that
$$x(n)\cos(\omega_0 n) \Longleftrightarrow \frac{1}{2} X(\omega - \omega_0) + \frac{1}{2} X(\omega + \omega_0) \tag{4.11}$$
and
$$x(n)\sin(\omega_0 n) \Longleftrightarrow \frac{1}{2j} X(\omega - \omega_0) - \frac{1}{2j} X(\omega + \omega_0) \tag{4.12}$$

Equations (4.11) and (4.12) can be readily derived by expressing $\sin(\omega_0 n)$ and $\cos(\omega_0 n)$ using Euler's formula and then applying the frequency-shifting property to each term.

Exercise 4.2.4

In this exercise, we will modulate the eight-sample signal
$$x(n) = u(n) - u(n - 8).$$
The DTFT of this signal was computed in part (d) of Exercise 4.1.2.

(a) Using the waveform generator function of PC-DSP accessed through menu selections *Sequences/Generate-sequence/Waveform-generator*, obtain the sequence
$$w(n) = \cos(0.2\pi n)$$
for the index range $n = 0, \ldots, 7$.

(b) Obtain the modulated sequence $y(n) = x(n)w(n)$. Compute its DTFT and then graph the magnitude and the phase. Observe the relationship between the transforms of $y(n)$ and $x(n)$.

(c) Repeat parts (a) and (b) with the signal

$$w(n) = \sin(0.2\pi n).$$

What are the differences in the spectrum when compared to the spectrum of part (b)? Pay particular attention to the phase of the transform.

Convolution Theorem

Given two arbitrary DTFT pairs

$$x(n) \Longleftrightarrow X(\omega)$$

and

$$h(n) \Longleftrightarrow H(\omega),$$

it can be shown that convolving the two signals in the time domain is equivalent to multiplying their transforms in the frequency domain; that is,

$$x(n) * h(n) \Longleftrightarrow X(\omega)H(\omega). \qquad (4.13)$$

The proof of (4.13) is instructive about the relationship between time- and frequency-domain methods and will be given here. It is possible to arrive at (4.13) by substituting the convolution equation given by (3.15) into the DTFT definition in (4.1) and manipulating the resulting expression. We will take a different approach and treat the DTFT as a decomposition of the signal. The inverse DTFT equation was given in (4.6) and is repeated here:

$$x(n) = \frac{1}{2\pi} \int_{-\pi}^{\pi} X(\omega)e^{j\omega n} d\omega. \qquad (4.14)$$

A method of numerically approximating an integral with a running sum was given in Exercise 3.4.4. Applying the rectangular approximation technique of that exercise to (4.14), we obtain

$$x(n) \approx \frac{\Delta\omega}{2\pi} \sum_{k=-M}^{M} X(k\Delta\omega)e^{jk\Delta\omega n}, \qquad (4.15)$$

where $\Delta\omega$ is the angular frequency increment. $\Delta\omega$ and M are chosen such that $M\,\Delta\omega = \pi$. The right side of the approximation in (4.15) becomes identical to the integral in (4.14) if the limit is taken for $\Delta\omega \to 0$.

One possible interpretation of (4.15) is as a decomposition of the signal $x(n)$ into $2M + 1$ complex exponential sequences at angular frequencies that are in the range $[-\pi, \pi]$ and that are also integer multiples of $\Delta\omega$. In the limit, $\Delta\omega$ approaches zero, and (4.14) becomes a decomposition of $x(n)$ into an *infinite* number of complex exponential sequences at *all* angular frequencies in the range $[-\pi, \pi]$.

If the signal $x(n)$ is used as input to a DTLTI system with system function $H(\omega)$, the response $y(n)$ of the system can be found using superposition. Recall that the response of the system to one complex exponential $e^{jk\Delta\omega n}$ is in the form $H(k\Delta\omega)e^{jk\Delta\omega n}$. The output signal $y(n)$ is

Sec. 4.2 Properties of the Discrete-time Fourier Transform

$$y(n) \approx \frac{\Delta\omega}{2\pi} \sum_{k=-M}^{M} H(k\Delta\omega)X(k\Delta\omega)e^{jk\Delta\omega n}. \quad (4.16)$$

In the limit, as $\Delta\omega$ approaches zero, the summation in (4.16) becomes an integral:

$$y(n) = \frac{1}{2\pi} \int_{-\pi}^{\pi} H(\omega)X(\omega)\, d\omega. \quad (4.17)$$

Hence

$$Y(\omega) = H(\omega)X(\omega) \quad (4.18)$$

Recall that in developing the time-domain convolution method of solving signal-system interaction problems in Chapter 3, we started with a decomposition of the input signal into scaled and shifted impulse sequences. Afterward, the superposition principle was used to find the response of the system. Equation (4.18) provides us with an alternative (frequency-domain) method of finding the system output. In this case, the input signal is decomposed into complex exponential sequences, and the superposition principle is applied.

Exercise 4.2.5

This exercise should be done without the use of the computer. Consider the following two sequences:

$$x(n) = \{1.0,\ 2.0,\ 3.0,\ 4.0,\ 5.0\},$$
$$\uparrow$$
$$h(n) = \{1.0,\ 1.0,\ 1.0\}.$$
$$\uparrow$$

Recall that the same two sequences were used in Exercise 3.2.2.

(a) Write the DTFT of each sequence as a polynomial of $e^{j\omega n}$. For example, the transform of $h(n)$ would be

$$H(\omega) = 1 + e^{j\omega} + e^{j2\omega}.$$

The transform of $x(n)$ can be written in a similar fashion.

(b) We know from (4.18) that convolving two sequences in the time domain is equivalent to multiplying their transforms. Obtain $Y(\omega)$ by multiplying the transforms $H(\omega)$ and $X(\omega)$. Note that you are performing polynomial multiplication. The product $Y(\omega)$ is another polynomial of $e^{j\omega n}$, and its coefficients correspond to samples of the sequence $y(n)$. Tabulate the nonzero range of $y(n)$. What is the length of the convolution result?

(c) Compare the resulting sequence $y(n)$ to that found in Exercise 3.2.2.

(d) Comment on how we can use the convolution function of PC-DSP to multiply two arbitrary polynomials.

Exercise 4.2.6

In this exercise, we will graphically verify the convolution theorem using two example signals. The signals for this exercise are on your distribution disks under the names EX4-2-6A.SEQ and EX4-2-6B.SEQ.

(a) Using PC-DSP, compute the DTFT of these two signals. Graph the magnitude and the phase of each spectrum.

(b) Compute the convolution of these signals using the convolution function accessed through menu selections *Operations/Processing-functions/Convolve-sequences*.

(c) On paper, roughly sketch the anticipated magnitude and phase characteristics of the convolution result based on (4.18).

(d) Compute the DTFT of the output sequence obtained in part (b). Graph the resulting spectrum, and compare it to your sketch.

(e) Tabulate the transforms of the two input sequences and the output sequence. Pick a particular value of the angular frequency ω. Verify the relationship given by (4.18), using the transform values at the frequency selected.

Differentiation in Frequency

For an arbitrary DTFT pair

$$x(n) \Longleftrightarrow X(\omega),$$

it can be shown that

$$nx(n) \Longleftrightarrow j\frac{dX(\omega)}{d\omega}. \tag{4.19}$$

This property is very useful in computing transforms of sequences with ramp-type components. Using (4.19) as the initial transform pair, and repeating the preceding property, we obtain

$$n^2 x(n) \Longleftrightarrow -\frac{d^2 X(\omega)}{d\omega^2}$$

and, in the general case,

$$n^k x(n) \Longleftrightarrow j^k \frac{d^k X(\omega)}{d\omega^k} \tag{4.20}$$

Parseval's Theorem

For an arbitrary transform pair

$$x(n) \Longleftrightarrow X(\omega),$$

the following equality holds:

$$\sum_{k=-\infty}^{\infty} |x(n)|^2 = \int_{-\pi}^{\pi} |X(\omega)|^2 \, d\omega. \tag{4.21}$$

But, as we have already established in Chapter 2, the left side of (4.21) represents the energy of the discrete-time signal $x(n)$. Therefore, the right side of (4.22) must be the frequency-domain representation of the signal energy.

4.3 SYMMETRY PROPERTIES OF THE DTFT

Some sequences have certain symmetry properties that can be exploited in computing the DTFT. In some cases, symmetry properties can be used for removing redundant parts of the transform. Consider a DTFT transform pair $x(n)$ and $X(\omega)$. Using the DTFT definition given by (4.1), it can easily be shown that

$$\text{DTFT}\{x^*(n)\} = X^*(-\omega) \qquad (4.22)$$

and
$$\text{DTFT}\{x^*(-n)\} = X^*(\omega). \qquad (4.23)$$

Equations (4.22) and (4.23) will be useful in deriving the symmetry properties of the DTFT.

Conjugate Symmetric Sequence

If the sequence $x(n)$ is conjugate symmetric, it satisfies the equation $x(n) = x^*(-n)$ for all values of the index n. Therefore, the transforms of these signals must also be equal; that is,

$$\text{DTFT}\{x(n)\} = \text{DTFT}\{x^*(-n)\},$$

which results in

$$X(\omega) = X^*(\omega) \qquad (4.24)$$

For (4.24) to be true, $X(\omega)$ must be real. Therefore, the transform of a conjugate symmetric sequence is real.

Conjugate Antisymmetric Sequence

A conjugate symmetric sequence $x(n)$ satisfies the equation $x(n) = -x^*(-n)$ for all values of the index n. Therefore, the transforms of these signals must also be equal; that is,

$$\text{DTFT}\{x(n)\} = \text{DTFT}\{-x^*(-n)\}$$
and
$$X(\omega) = -X^*(\omega) \qquad (4.25)$$

This requires $X(\omega)$ to be purely imaginary. Thus, the transform of a conjugate antisymmetric sequence is purely imaginary.

Exercise 4.3.1

The goal of this exercise will be to verify the preceding properties using example sequences.

(a) Generate a 100-sample complex sequence $x(n)$ with arbitrary sample values. You might want to use the waveform generator or the formula-entry method for this purpose.

(b) Compute the DTFT of this sequence. Tabulate and graph the result.

(c) Compute the sequences $x_E(n)$ and $x_O(n)$, the conjugate symmetric and conjugate antisymmetric components of $x(n)$, respectively. Compute the DTFT of each of these components. Tabulate the results.

(d) Is the DTFT of $x_E(n)$ purely real? Is the DTFT of $x_O(n)$ purely imaginary? Does the sum of the two transforms equal the transform of $x(n)$?

Real Sequence

For a real sequence $x(n)$, the equation $x(n) = x^*(n)$ is satisfied for all n; therefore,

$$X(\omega) = X^*(-\omega), \qquad (4.26)$$

which implies that the transform $X(\omega)$ is conjugate symmetric. This property is very important since, in real-time signal-processing applications, we mostly deal with real-valued sequences. If we write the transform $X(\omega)$ in polar complex form as

$$X(\omega) = |X(\omega)|e^{j\,\arg[X(\omega)]},$$

then (4.26) implies that

$$|X(\omega)| = |X(-\omega)|, \qquad (4.27)$$

$$\arg[X(\omega)] = -\arg[X(-\omega)]. \qquad (4.28)$$

The magnitude of the transform is an even function of ω, and its phase is odd. We know from previous discussion that the DTFT of a sequence is a periodic function of ω with a period of 2π radians, and a computed transform only needs to be shown in the range $-\pi \leq \omega \leq \pi$. If the sequence transformed is real, then the left half of the transform for $-\pi \leq \omega < 0$ is dependent on its right half for $0 \leq \omega \leq \pi$, and therefore only the right half needs to be shown. When we discuss the design of discrete-time filters with real-valued impulse responses in Chapters 7 and 8, we will specify the desired frequency-domain behavior only for the interval $0 \leq \omega \leq \pi$.

Exercise 4.3.2

(a) Generate a 100-sample real sequence $x(n)$ with arbitrary sample values. You might want to use the waveform generator or the formula-entry method for this purpose.
(b) Compute the DTFT of this sequence. Graph the magnitude and the phase of the transform. Observe the symmetry properties of the magnitude and the phase spectra. Do they agree with (4.27) and (4.28)?
(c) Tabulate the transform computed in part (b). Pick a particular value ω_1 for the angular frequency. Observe the magnitude and the phase at frequencies ω_1 and $-\omega_1$. Comment.

Imaginary Sequence

If a sequence $x(n)$ is purely imaginary, the property $x(n) = -x^*(n)$ holds for all n, and therefore

$$X(\omega) = -X^*(-\omega), \qquad (4.29)$$

which implies that the transform $X(\omega)$ is conjugate antisymmetric. As a result of this, the following equations can be written for real and imaginary parts of the transform:

$$\text{Re}\{X(\omega)\} = -\text{Re}\{X(-\omega)\}, \quad (4.30)$$

$$\text{Im}\{X(\omega)\} = \text{Im}\{X(-\omega)\}. \quad (4.31)$$

In this case, the real part of the transform is an odd function of ω, and its imaginary part is even.

Exercise 4.3.3

(a) Generate a 100-sample imaginary sequence $x(n)$ with arbitrary sample values. An easy method of doing this would be to multiply the real sequence obtained in the previous exercise by $(0 + j1)$.
(b) Compute the DTFT of this sequence. Graph the real and imaginary parts of the transform. Observe the symmetry properties. Do they agree with (4.30) and (4.31)?
(c) Tabulate the transform computed in part (b). Pick a particular value ω_1 for the angular frequency. Observe real and imaginary parts of the transform at frequencies ω_1 and $-\omega_1$. Comment.

4.4 FREQUENCY-DOMAIN ANALYSIS OF SYSTEMS

In this section, some basic concepts that are derived from the system function concept will be introduced. In Chapter 3, the system function $H(\omega)$ was defined as the DTFT of the impulse response $h(n)$; that is,

$$H(\omega) = \sum_{n=-\infty}^{\infty} h(n) e^{-j\omega n}$$

In general, $H(\omega)$ is a complex function and can be expressed in either Cartesian or polar complex forms. The magnitude of the system function is sometimes referred to as the *magnitude response* of the system. Similarly, the phase of the system function is sometimes called the *phase response*. The decibel magnitude response of a system is obtained by

$$|H(\omega)|_{dB} = 20 \log_{10}[|H(\omega)|] \quad (4.32)$$

This is also referred to as the *gain characteristic* of the system. Its negative is called the *attenuation characteristic*. The *time delay* of a system is defined as

$$\tau_d(\omega) = -\frac{\arg[H(\omega)]}{\omega}, \quad (4.33)$$

and the *group delay* is defined as

$$\tau_g(\omega) = -\frac{\partial}{\partial \omega}\{\arg[H(\omega)]\}. \quad (4.34)$$

The time delay characteristic of a system provides information about how much each frequency component of the input signal is delayed by the system. Recall that any discrete-time signal $x(n)$ can be written as a linear combination of sinusoidal basis functions. If basis functions at different angular frequencies are delayed by different amounts, the output signal obtained by combining these will be a distorted version of the input signal. This effect is known as *phase distortion*. In some applications, the time delay is desired to be a constant, independent of frequency so that phase distortion can be avoided:

$$\tau_d(\omega) = C = \text{constant}.$$

This, in turn, requires

$$\arg[H(\omega)] = -C\omega.$$

Thus, for constant time delay, the phase response of the system must be a linear function of the angular frequency ω. For this reason, linear-phase systems are very important in signal-processing applications. The properties of linear-phase systems will be discussed in more detail in the next section.

4.5 LINEAR-PHASE SYSTEMS

It can be shown that a linear-phase characteristic is only possible for finite impulse response (FIR) systems. A length-N FIR filter is typically described by means of its impulse response $h(n)$ in the range of the sample index $0 \leq n \leq N - 1$. (The impulse response is zero valued outside this range.) There are four types of linear-phase FIR filters:

Type 1. The filter length N is odd, and the impulse response satisfies

$$h(n) = h(N - 1 - n), \quad n = 0, \ldots, N - 1. \quad (4.35a)$$

Type 2. The filter length N is even, and the impulse response satisfies

$$h(n) = h(N - 1 - n), \quad n = 0, \ldots, N - 1. \quad (4.35b)$$

Type 3. The filter length N is odd, and the impulse response satisfies

$$h(n) = -h(N - 1 - n), \quad n = 0, \ldots, N - 1. \quad (4.35c)$$

Type 4. The filter length N is even, and the impulse response satisfies

$$h(n) = -h(N - 1 - n), \quad n = 0, \ldots, N - 1. \quad (4.35d)$$

The system function for an FIR filter is

$$H(\omega) = \sum_{n=0}^{N-1} h(n)e^{-j\omega n}, \quad (4.36)$$

where the limits on the DTFT summation are changed to 0 and $N - 1$ since the filter impulse response is zero outside this range. Let $h(n)$ be a type-1 filter. Since N is an odd number, (4.36) can be written as

$$H(\omega) = \sum_{n=0}^{M-1} h(n)e^{-j\omega n} + \sum_{n=M+1}^{N-1} h(n)e^{-j\omega n} + h(M), \quad (4.37)$$

Sec. 4.5 Linear-phase Systems

where we have used the convention

$$M = \frac{N-1}{2}.$$

Applying a variable change to the second summation in (4.37), we obtain

$$H(\omega) = \sum_{n=0}^{M-1} h(n)e^{-j\omega n} + \sum_{n=0}^{M-1} h(N-1-n)e^{-j\omega(N-1)}e^{j\omega n} + e^{-j\omega M}h(M),$$

which can also be written as

$$H(\omega) = e^{-j\omega M} \sum_{n=0}^{M-1} \{h(n)e^{j\omega M}e^{-j\omega n} + h(N-1-n)e^{-j\omega M}e^{j\omega n}\} + e^{-j\omega M}h(M).$$

This expression can be simplified using Euler's formula and the symmetry property given by (4.35a), yielding

$$H(\omega) = e^{-j\omega M} \sum_{n=0}^{M-1} \{2h(n)\cos[\omega(n-M)]\} + h(M). \qquad (4.38)$$

The system function in (4.38) is in the form

$$H(\omega) = A(\omega)e^{j\Theta(\omega)}, \qquad (4.39)$$

where the phase is clearly a linear function of the angular frequency; that is,

$$\Theta(\omega) = -\omega M.$$

System functions for other types of linear-phase systems can be derived similarly and will be left to the exercises.

Exercise 4.5.1

Consider an FIR filter with the impulse response

$$h(n) = \{1, -8, 27, -54, 67, -54, 27, -8, 1\}.$$
$$\uparrow$$

(a) This is a type 1 linear-phase filter. On paper, show that the analytical form of the system function is given by (4.39) with

$$A(\omega) = 2\cos(4\omega) - 16\cos(3\omega) + 54\cos(2\omega) - 108\cos(\omega) + 67$$

and

$$\Theta(\omega) = -4\omega.$$

Roughly sketch the amplitude and the phase in the angular frequency range $(0, 2\pi)$.

(b) Using PC-DSP, generate the impulse response of this filter. You may want to use the menu selections *Sequences/Generate-sequence/Read-from-keyboard* and enter desired sample values. Compute the system function using DTFT. Graph the magnitude and the phase of the system function.

(c) Compare the screen plots obtained in part (b) to your hand sketches. Pay special attention to the phase characteristic. Is it linear? Note that the broken-line appearance of the phase is due to additions and subtractions of 2π radians to keep the phase in the range $(0, 2\pi)$.

Exercise 4.5.2

Consider an FIR filter with the impulse response
$$h(n) = \{1, -5, 6, -4, -6, -4, 6, -5, 1\}.$$
$$\uparrow$$

(a) This is also a type 1 linear-phase filter. As in the previous exercise, derive the analytical forms of amplitude and phase characteristics. Afterward, roughly sketch the amplitude and the phase in the angular frequency range $(0, 2\pi)$.

(b) Using PC-DSP, generate the impulse response of this filter. Compute the system function using DTFT. Graph the magnitude and the phase of the system function.

(c) Compare the screen plots obtained in part (b) to your hand sketches. Note that, in this case, the magnitude response graphed does not match your hand sketch exactly. The *magnitude* characteristic is nonnegative by definition, whereas the *amplitude* characteristic might be negative in some frequency ranges. The system function displayed on the screen is in the form
$$H(\omega) = |H(\omega)|e^{j\,\arg[H(\omega)]},$$
rather than the form given by (4.39). The phase functions $\Theta(\omega)$ and $\arg[H(\omega)]$ are identical at frequencies where $A(\omega)$ is positive, and they differ by π radians at frequencies where $A(\omega)$ is negative.

Exercise 4.5.3

(a) Consider again the FIR filter used in Exercise 4.5.1. Change the signs of the last four samples of the impulse response to obtain a type-3 linear-phase filter. On paper, show that the analytical form of the system function is given by (4.39) with
$$A(\omega) = 2\sin(4\omega) - 16\sin(3\omega) + 54\sin(2\omega) - 108\sin(\omega) + 67$$
and
$$\Theta(\omega) = \pi - 4\omega.$$
Roughly sketch the amplitude and the phase in the angular frequency range $(0, 2\pi)$.

(b) Using PC-DSP, generate the impulse response of this filter. Compute the system function using DTFT. Graph the magnitude and the phase of the system function. Compare the screen plots obtained in part (b) to your hand sketches.

4.6 DISCRETE FOURIER TRANSFORM

Frequency-domain representation of discrete-time signals with the discrete-time Fourier transform (DTFT) was discussed in the previous sections of this chapter. The DTFT is general enough to be applied to both finite-length and infinite-length signals (provided, of course, that they satisfy the existence conditions). For a given discrete-time signal $x(n)$, the DTFT is a continuous function of the angular frequency ω. In digital computer applications, a transform technique that allows a discrete-time signal to be represented with a discrete transform would be preferred. This provides part of the motivation for using the discrete Fourier transform (DFT). Also, very efficient numerical techniques exist for computing the DFT, making it very attractive for signal-processing applications.

Consider a finite-length discrete-time signal $x(n)$ that is nonzero only in the range $0 \leq n \leq N - 1$. The DFT of $x(n)$ is defined as

Sec. 4.6 Discrete Fourier Transform

$$X(k) = \sum_{n=0}^{N-1} x(n) e^{-j2\pi nk/N}, \qquad k = 0, \ldots, N-1. \tag{4.40}$$

Note that a N-point sequence leads to an N-point transform. Given the transform $X(k)$, the signal $x(n)$ can be found using the inverse DFT equation

$$x(n) = \frac{1}{N} \sum_{k=0}^{N-1} X(k) e^{j2\pi nk/N}, \qquad n = 0, \ldots, N-1. \tag{4.41}$$

Together, (4.40) and (4.41) form a transform pair. The similarity between DTFT and DFT definitions is important enough to warrant further exploration. The DTFT of a signal $x(n)$ was defined by (4.1), which will be repeated here for convenience:

$$X(\omega) = \sum_{n=-\infty}^{\infty} x(n) e^{-j\omega n}. \tag{4.42}$$

If $x(n)$ is a finite-length signal that is nonzero only in the range $0 \le n \le N-1$, the limits of the summation in (4.42) can be changed without affecting the result; that is,

$$X(\omega) = \sum_{n=0}^{N-1} x(n) e^{-j\omega n}. \tag{4.43}$$

Comparing (4.43) with the DTFT definition given by (4.40), it is clear that

$$X(k) = X(\omega)\big|_{\omega = 2\pi k/N}; \tag{4.44}$$

that is, the DFT of a finite-length signal corresponds to its DTFT sampled at N equally spaced angular frequencies in the interval $(0, 2\pi)$.

Exercise 4.6.1

(a) Using PC-DSP, generate samples of the signal

$$x(n) = u(n) - u(n - 32)$$

for $n = 0, \ldots, 31$. Compute the DTFT of this signal. Graph the magnitude and the phase of the DTFT.

(b) Compute the DFT of the signal $x(n)$ using the menu selections *Transforms/DFT*. In the data-entry window presented, enter the names of the input and output sequences. Specify the DFT size as 32 (more on this later). Select the rectangular window. (The signal is optionally multiplied with a window function before the DFT is computed; however, multiplication with a rectangular window is equivalent to not using a window function.) Graph the DFT result.

(c) Tabulate DTFT and DFT results side by side. Compare the two in light of Eq. (4.44). For example, sample number $k = 5$ of the DFT should be equal to the DTFT at the frequency $\omega_5 = 10\pi/32$.

Equation (4.44) describes the relationship between the DFT and the DTFT of a finite-length sequence. The DFT turns out to be equal to the DTFT at N equally spaced frequencies. On the other hand, the N-sample DFT result uniquely represents the N-sample sequence $x(n)$, since the latter can be obtained from the former using the inverse DFT equation given by (4.41). This means that the DTFT of a finite-length sequence has some redundancy in it, since we only need to know it at N equally spaced frequencies to find the sequence $x(n)$. As a result, we should be able to obtain the DTFT (a continuous function of ω) of an N-sample sequence from the knowledge of its N-sample DFT.

Substituting (4.41) into (4.43), we obtain

$$X(\omega) = \sum_{n=0}^{N-1} \left[\frac{1}{N} \sum_{k=0}^{N-1} X(k) e^{j2\pi nk/N} \right] e^{-j\omega n}.$$

Interchanging the two summations and rearranging yields

$$X(\omega) = \frac{1}{N} \sum_{k=0}^{N-1} X(k) \sum_{n=0}^{N-1} e^{j2\pi nk/N} e^{-j\omega n}$$

or, equivalently,

$$X(\omega) = \frac{1}{N} \sum_{k=0}^{N-1} X(k) \left[\frac{1 - e^{-jN(\omega - 2\pi k/N)}}{1 - e^{-j(\omega - 2\pi k/N)}} \right]. \tag{4.45}$$

This represents an interpolation formula in which the DTFT is obtained by interpolating between samples of the DFT. Note that this is only valid when the sequence $x(n)$ is of finite length.

Zero Padding

As a side note, we will explore the effects of zero padding on the DFT. *Zero padding* is the act of extending a signal by appending zero-valued samples to it. Given an N-sample signal $x(n)$, let a new signal $\tilde{x}(n)$ be defined as

$$\tilde{x}(n) = \begin{cases} x(n), & n = 0, \ldots, N-1 \\ 0, & n = N, \ldots, M-1, \end{cases}$$

where $M > N$. The M-sample DFT of this signal is

$$X(k) = \sum_{n=0}^{M-1} \tilde{x}(n) e^{-j2\pi nk/M}. \tag{4.46}$$

Realizing that the last $M - N$ terms in the summation of (4.46) are zero valued, we can write

$$X(k) = \sum_{n=0}^{N-1} x(n) e^{-j2\pi nk/M}. \tag{4.47}$$

Note that this is equivalent to evaluating the DTFT of $x(n)$ at M equally spaced frequencies in the interval $(0, 2\pi)$, which is evident by a comparison of (4.43) and (4.47). Thus, extending the length of a signal by zero padding and then computing its DFT with the new length corresponds to sampling the DTFT of the signal at more points. The DFT of the zero-padded signal does not contain any new information, however, since the additional samples could have also been obtained simply by interpolation as dictated by (4.45).

Exercise 4.6.2

In Exercise 4.6.1, the DFT of the 32-sample signal

$$x(n) = u(n) - u(n - 32)$$

Sec. 4.6 Discrete Fourier Transform

was computed using PC-DSP. In this exercise, we will pad the signal with zeros and then compute the DFT.

(a) Use the menu selections *Transforms/DFT* to access the DFT function. In the data-entry window presented, enter 64 for the DFT size parameter. Since the original signal has 32 samples, entering a larger value for the DFT size causes the signal to be padded with zeros. The transform sequence obtained should have 64 samples corresponding to the DFT of $\tilde{x}(n)$, that is, $x(n)$ padded with 32 zero-amplitude samples. Tabulate the 64-point transform obtained side by side with the 32-point transform of Exercise 4.6.1. The larger transform amounts to evaluating the DTFT of the signal at twice as many frequencies. Observing tabulated results, verify the following:

$$\tilde{X}(2k) = X(k), \quad k = 0, \ldots, 31.$$

(b) Compare various even- and odd-indexed samples of the 64-point DFT result to appropriate values in the DTFT of the 32-sample signal $x(n)$. [Recall that the DTFT was computed in part (a) of Exercise 4.6.1.]

(c) Repeat parts (a) and (b) with a DFT size of 80. Which samples of the 32- and 80-point DFT results are identical in this case?

Infinite-length Sequence

Now consider an infinite-length sequence $x(n)$. Its DTFT is given by (4.1) and (4.42). By sampling the DTFT at N equally spaced frequencies in the interval $(0, 2\pi)$, we obtain

$$X_1(k) = \sum_{n=-\infty}^{\infty} x(n) e^{-j2\pi kn/N}, \quad k = 0, \ldots, N - 1.$$

It is possible to think of $X_1(k)$ as the DFT of some N-sample sequence $x_1(n)$; that is,

$$X_1(k) = \sum_{n=0}^{N-1} x_1(n) e^{-j2\pi kn/N}.$$

It can be shown that the infinite-length sequence $x(n)$ and the N-sample sequence $x_1(n)$ are related by

$$x_1(n) = \sum_{r=-\infty}^{\infty} x(n - rN); \quad (4.48)$$

that is, $x_1(n)$ is obtained by aliasing $x(n)$ with its shifted versions. This interesting result will be particularly important later when the *circular convolution* issue is discussed.

Exercise 4.6.3

Consider the 11-sample ramp sequence

$$x(n) = n\,u(n), \quad n = 0, \ldots, 10.$$

(a) Generate the sequence $x(n)$ using either the waveform generator function or the formula-entry function of PC-DSP. Compute its DTFT and tabulate the result.

(b) Obtain an 8-point transform sequence $X_1(k)$ by sampling the DTFT at eight equally spaced frequencies. Note that the DTFT result is tabulated as a function of the angular frequency. You will first need to determine the eight frequency values at which the DTFT will be sampled and then look up the transform at those frequencies. The easiest method of doing this in PC-DSP is to bring up the DTFT table and the text editor side by side so that you can read values from the table and type into the text editor. When finished, save your text file and import it into PC-DSP using the menu selections *Sequences/Import-ASCII-Data-File*.

(c) Let $x_1(n)$ be the 8-point inverse DFT $X_1(k)$. Without computer aid, compute $x_1(n)$ using Eq. (4.48).

(d) Using PC-DSP, compute the sequence $x_1(n)$ as the 8-point inverse DFT of $X_1(k)$. Tabulate the result and compare to the original ramp sequence $x(n)$. How many samples match between the two and in which range of the sample index n?

Note: The sequence $x_1(n)$ should be real since the transform samples are conjugate symmetric; however, a negligibly small imaginary part may result from machine round-off errors.

4.7 PROPERTIES OF THE DISCRETE FOURIER TRANSFORM

In this section, we will discuss some of the important properties of the discrete Fourier transform. The tone of presentation will be similar to that of Section 4.2, and some of the properties will seem trivial; however, they are important for effective use of the DFT in signal processing. To keep the notation simple, we will denote the forward transform with

$$X(k) = \text{DFT}\{x(n)\}$$

and the inverse transform with

$$x(n) = \text{DFT}^{-1}\{X(k)\}.$$

The notation

$$x(n) \Longleftrightarrow X(k)$$

will be employed to indicate that $x(n)$ and $X(k)$ are a DFT transform pair. Proofs of some of these properties will be given, and some will be left to the reader.

Linearity

The DFT is a linear transform. Given two DFT pairs

$$x_1(n) \Longleftrightarrow X_1(k)$$

and

$$x_2(n) \Longleftrightarrow X_2(k)$$

and two constants α and β, it can be shown that

$$\alpha x_1(n) + \beta x_2(n) \Longleftrightarrow \alpha X_1(k) + \beta X_2(k). \tag{4.49}$$

It is assumed that the two sequences have the same length N.

Exercise 4.7.1

In this exercise, we will numerically verify the linearity property of the DFT.

(a) Using the formula-entry method of PC-DSP accessed with menu selections *Sequences/Generate-sequence/Formula-entry-method*, generate the following 25-sample signals:

Sec. 4.7 Properties of the Discrete Fourier Transform

$$x_1(n) = e^{-0.1n}, \quad n = 0, \ldots, 24$$
$$x_2(n) = n\sin(0.3n), \quad n = 0, \ldots, 24.$$

Compute the DFT of each signal. The DFT function of PC-DSP is accessed with the menu selections *Transforms/DFT*. In the data-entry window presented, enter the names of input and transform sequences. Select the *rectangular* window (more on this later) and specify the DFT size as 25. Tabulate and graph the results.

(b) Use PC-DSP to obtain the sum of these two signals; that is, $x(n) = x_1(n) + x_2(n)$. Compute the DFT of $x(n)$. Tabulate it along with the individual transform of $x_1(n)$ and $x_2(n)$.

(c) Pick a few arbitrary values for k and show for each that the transform of the sum of two sequences is equal to the sum of two transforms; that is,

$$X(k) = X_1(k) + X_2(k).$$

(d) Obtain a new sequence $x_3(n)$ by scaling $x_1(n)$ as

$$x_3(n) = 5x_1(n).$$

Compute the DFT of $x_3(n)$. Tabulate it along with the transform of $x_1(n)$. Pick a few arbitrary values for k and show for each that the transforms satisfy

$$X_3(k) = 5X_1(k).$$

Circular Shifting

Consider an arbitrary DFT pair

$$x(n) \Longleftrightarrow X(k),$$

where $x(n)$ is an N-sample sequence and $X(k)$ is its N-point DFT. Circularly shifting the signal $x(n)$ in the time domain is equivalent to multiplying its transform by a complex exponential; that is,

$$x((n - M))_N \Longleftrightarrow e^{-j2\pi kM/N} X(k). \tag{4.50}$$

The notation $x((n - M))_N$ indicates a circular shift of M samples. This is a fundamental property of the DFT and is illustrated in Fig. 4.1.

If we define a sequence $x_1(n)$ through the transform relationship

$$x_1(n) \Longleftrightarrow e^{-j2\pi kM/N} X(k),$$

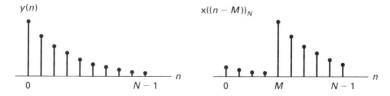

Figure 4.1 Illustration of circular shifting.

then it can be shown that

$$x_1(n) = \begin{cases} x(n - M + N), & n = 0, \ldots, M - 1 \\ x(n - M), & n = M, \ldots, N - 1, \end{cases} \quad (4.51)$$

which corresponds to a circular shift of M samples applied to the sequence $x(n)$.

Exercise 4.7.2

(a) Using PC-DSP, generate the 25-sample sequence

$$x(n) = 24u(n) - n\, u(n), \quad n = 0, \ldots, 24.$$

Use either the waveform generator function or the formula-entry method. Afterward, compute the 25-point DFT of this sequence.

(b) Multiply the DFT result with the complex exponential sequence

$$X_1(k) = e^{-j8\pi k/25} X(k)$$

and compute the inverse DFT of $X_1(k)$ to obtain $x_1(n)$. Remember to remove any imaginary parts that are due to machine round-off errors. Compare the sequences $x(n)$ and $x_1(n)$. Verify the relationship given by (4.51) between the two sequences.

(c) Repeat part (b) with

$$X_2(k) = e^{-j14\pi k/25} X(k).$$

Circular Convolution

Consider two DFT pairs given by

$$x(n) \iff X(k),$$
$$h(n) \iff H(k),$$

where both $x(n)$ and $h(n)$ are assumed to be N-sample sequences that exist for $0 \le n \le N - 1$. It can be shown that

$$x(n) \otimes h(n) \iff X(k)H(k). \quad (4.52)$$

The symbol $\otimes$ indicates circular convolution, which is defined as

$$x(n) \otimes h(n) = \sum_{n=0}^{N-1} x(m) h((n - m))_N. \quad (4.53)$$

Exercise 4.7.3

(a) Using PC-DSP, generate the following 10-sample sequences:

$$x(n) = \begin{cases} 1, & n = 0, \ldots, 5 \\ 0, & n = 6, \ldots, 9, \end{cases}$$

$$h(n) = \begin{cases} 1, & n = 0, \ldots, 7 \\ 0, & n = 8, \ldots, 9. \end{cases}$$

(b) On paper, compute the 10-point circular convolution of these two sequences.

(c) Use PC-DSP to compute the circular convolution. Note that PC-DSP does not have a built-in

Sec. 4.7 Properties of the Discrete Fourier Transform

function for circular convolution. Forward and inverse DFT functions must be used instead. First, compute the 10-point transforms $X(k)$ and $H(k)$ using menu selections *Transforms/DFT*. Afterward, multiply the two transforms with menu selections *Operations/Arithmetic-operations/ Multiply-sequences*, and, finally, compute the inverse DFT of the product with menu selections *Transforms/Inverse-DFT*. Compare the result to that obtained manually in part (b).

Even though multiplication of two DFTs corresponds to a circular convolution of the corresponding sequences in the time domain, it is possible to use the DFT for computing linear convolution. To do this, we need to explore the relationship between linear and circular convolution operations. Let's make the following definitions:

$$y_l(n) = x(n) * h(n) \quad \text{(linear convolution)},$$
$$y_c(n) = x(n) \otimes h(n) \quad \text{(circular convolution)}.$$

The linear convolution result is $2N - 1$ samples long, whereas the circular convolution result has N samples. The two sequences have the time-domain aliasing relationship given by (4.48); that is,

$$y_c(n) = \sum_{r=-\infty}^{\infty} y_l(n - rN). \tag{4.54}$$

Exercise 4.7.4

In this exercise, the emphasis will be on using the DFT as a tool for computing the linear convolution of two sequences. Consider the following two sequences:

$$x_1(n) = \{3, 4.2, 1, 1, 0, 7, -1, 0, 2\},$$
$$\qquad\qquad\uparrow$$
$$x_2(n) = \{1.2, 3, 0, -0.5, 2\}.$$
$$\qquad\uparrow$$

(a) First, compute the linear convolution of the sequences $x_1(n)$ and $x_2(n)$ using the *Convolution* function of PC-DSP. What is the length of the result?

(b) In some cases, it may be computationally more efficient to use the discrete Fourier transform to compute the convolution of two sequences. Initially, we will attempt to do it in a way that a not so careful student might do. The sequence $x_2(n)$ can be extended to 9 points by zero padding so that both sequences have the same length. (In PC-DSP, this is achieved by specifying a DFT size that is greater than the length of the sequence.) Compute 9-point DFTs $X_1(k)$ and $X_2(k)$ of the signals $x_1(n)$ and $x_2(n)$, respectively:

$$X_1(k) = \text{DFT}\{x_1(n)\}$$

and

$$X_2(k) = \text{DFT}\{x_2(n)\}.$$

Multiply $X_1(k)$ and $X_2(k)$ and then compute the inverse DFT of the product. Take the real part of the result to remove the imaginary part that is due to round-off error. How does this result compare to the correct linear convolution sequence obtained in part (a)? How many samples of the result are correct as the linear convolution of the two sequences? Why?

(c) What is the minimum DFT size that should be used if we are to obtain the true linear convolution of $x_1(n)$ and $x_2(n)$? Extend both sequences to the minimum required DFT size before computing the transforms. Repeat part (b).

TABLE 4.3 Some Properties of the DFT

Signal		Transform
1. $x(n)$	$\Longleftrightarrow$	$X(k)$
2. $x^*(n)$	$\Longleftrightarrow$	$X^*((-k))_N$
3. Linearity		
$\alpha x_1(n) + \beta x_2(n)$	$\Longleftrightarrow$	$\alpha X_1(k) + \beta X_2(k)$, all α and β
4. Circular shifting		
$x((n-M))_N$	$\Longleftrightarrow$	$e^{-j2\pi kM/N} X(k)$
5. Circular convolution		
$x(n) \otimes h(n)$	$\Longleftrightarrow$	$X(k)H(k)$
6. Conjugate symmetric sequence		
$x(n) = x^*((-n))_N$	$\Longleftrightarrow$	$X(k)$ real
7. Conjugate antisymmetric sequence		
$x(n) = -x^*((-n))_N$	$\Longleftrightarrow$	$X(k)$ imaginary
8. $x(n)$ real	$\Longleftrightarrow$	$X(k) = X^*((-k))_N$
		$\|X(k)\| = \|X((-k))_N\|$
		$\arg[X(k)] = -\arg[X((-k))_N]$
9. $x(n)$ imaginary	$\Longleftrightarrow$	$X(k) = -X^*((-k))_N$

(d) Pad $x_1(n)$ and $x_2(n)$ with five more zeros so that their lengths exceed the minimum DFT length required for linear convolution by five samples. Repeat part (c) under these circumstances. How does the excess length affect the result?

Exercise 4.7.5

In this exercise, we will use some of the results of Exercise 4.7.4.

(a) Using the linear convolution result obtained in part (a) of Exercise 4.7.4 with the time-domain aliasing equation given by (4.54), construct the 9-point circular convolution result. Note that, in this particular case, this amounts to computing

$$y_c(n) = y_l(n) + y_l(n+9), \quad n = 0, \ldots, 8.$$

In PC-DSP, use linear shift and addition operations, and then truncate the result into the desired range of the index n using menu selections *Sequences/Sequence-editing/Copy-sequence*. Compare the resulting nine-sample sequence to that obtained in part (b) of Exercise 4.7.4.

(b) Repeat part (a) using the extended DFT size that was used in part (c) of Exercise 4.7.4. Comment on the result.

Table 4.3 lists some of the properties of the discrete Fourier transform. The sequences involved are assumed to start at index 0 and end at index $N-1$.

Exercise 4.7.6

(a) Using PC-DSP, generate a 64-sample complex sequence $x(n)$ with arbitrary sample values. You might want to use the waveform generator or the formula-entry method for this purpose.

(b) Compute the 64-point DFT of this sequence. Tabulate and graph the result.

(c) Compute the length-64 sequences $x_{ep}(n)$ and $x_{op}(n)$ defined as

$$x_{ep}(n) = \frac{1}{2}\{x(n) + x^*((-n))_N\}, \quad n = 0, \ldots, N-1$$

Sec. 4.7 Properties of the Discrete Fourier Transform

$$x_{op}(n) = \frac{1}{2}\{x(n) - x^*((-n))_N\}, \quad n = 0, \ldots, N - 1,$$

where $N = 64$. These are the periodic conjugate-symmetric and periodic conjugate-antisymmetric components of $x(n)$, respectively, and can be computed using menu selections *Operations/Arithmetic-operations/Even-component* and *Operations/Arithmetic-operations/Odd-component*. In the data-entry window presented, select *Conjugate* and *Modulo-N* options.

(d) Compute the DFT of each of these components. Tabulate the results simultaneously with the transform obtained in part (b). Verify that

$$\text{DFT}\{x_{ep}(n)\} = \text{Re}\{X(k)\},$$
$$\text{DFT}\{x_{op}(n)\} = \text{Im}\{X(k)\}.$$

Exercise 4.7.7

(a) Using PC-DSP, generate a 100-sample real sequence $x(n)$ with arbitrary sample values. You might want to use the waveform generator or the formula-entry method for this purpose.

(b) Compute the DFT of this sequence. Graph the magnitude and the phase of the transform. Observe the symmetry properties of the magnitude and the phase spectra and compare to entry 8 of Table 4.3.

(c) Tabulate the transform computed in part (b). Pick a particular value of the transform index k. Observe the magnitude and the phase of the DFT at index values k and $100 - k$. Comment.

5

Sampling and Reconstruction

> Everything that can be invented has been invented. —Charles H. Duell, Commissioner, U.S. Patent Office, 1899

Most signals that we confront in our daily lives are analog signals. Some examples are speech and music signals, biomedical signals, seismic signals, and most communications signals that are broadcast through the air. For processing through a digital system, an analog signal must first be converted to a digital signal. This process is referred to as *analog-to-digital conversion*, and the system that performs it is called an *analog-to-digital converter*, or *A/D converter* for short. The inverse process, of converting a digital signal to an analog signal is referred to as *digital-to-analog conversion*, and the system that performs it is called a *digital-to-analog converter*, or *D/A converter* for short. Figure 5.1 illustrates a typical situation in which A/D and D/A converters are used in conjunction with a digital system to facilitate digital processing of an analog signal.

Analog-to-digital conversion can be thought of as the act of measuring an analog signal at periodic intervals and expressing the result of each measurement in a form that can be used by a digital computer. For example, the amplitude of the voltage at the output of an audio amplifier could be measured several thousand times each second, and the measurement results could be expressed in binary form to represent the analog audio signal. Intuitively, we might expect the quality of this representation to be dependent on

Chap. 5 Sampling and Reconstruction

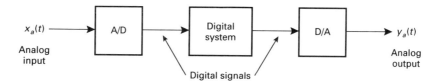

Figure 5.1 Digital processing of an analog signal.

how often the measurements are taken and how accurate the corresponding binary numbers are.

The actual physical implementation of an A/D converter depends on a number of factors, such as the type of analog input signal (voltage, current, vibration, temperature, or the like), the type of digital system used, the rate at which measurements are taken, and the method of representing numbers in binary form. To analyze the conversion process, however, we will adopt the mathematical model shown in Fig. 5.2. This A/D converter model consists of two major blocks: an ideal sampler and an encoder. The ideal sampler converts the analog signal $x_a(t)$ into a discrete-time sequence $x(n)$. The task of the encoder is to code the amplitude of each sample of $x(n)$ into a digital form. Within this block, sample values are first quantized into a finite number of discrete amplitude levels. Afterward, these discrete amplitude levels are expressed in binary form. For example, on a particular system, a sample with amplitude 0.75 might be coded into the binary word 0110000000000000. Finally, the binary word might be transmitted to the target processor using, for example, 3-V pulses for 1s, and -3-V pulses for 0s.

The practical utility of this A/D converter model lies in the fact that the first component, the ideal sampler, is identical for all A/D converters modeled using this approach. All hardware-dependent features of an A/D converter are modeled within the second component, the encoder block. Thus, the use of this model allows us to study the fundamental principles of analog to discrete-time conversion (as opposed to analog-to-digital conversion) without having to pay attention to hardware-specific implementation issues. In this chapter, we will mainly concentrate on the ideal sampler component of Fig. 5.2. The quantization function of the encoder will be briefly treated in Section 5.3.

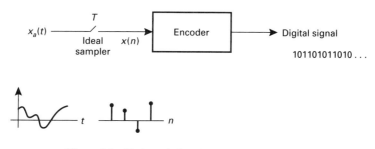

Figure 5.2 Mathematical model for an A/D converter.

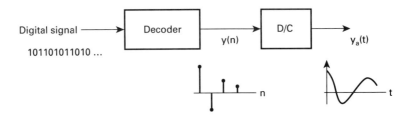

Figure 5.3 Mathematical model for a D/A converter.

A similar mathematical model can be developed for a D/A converter. (See Fig. 5.3.) This model also consists of two components. The first block, the decoder, converts the digital signal to a discrete-time sequence. This part of the system encapsulates all operations in a D/A converter that are hardware dependent, for example, the specific binary numbering scheme of the system and any quantization-related operations. The second block converts the discrete-time sequence to an analog signal. For the lack of a better term, we will call this block a *discrete-to-continuous converter*.

Using the mathematical models given for A/D and D/A converters, a digital system that processes analog signals can be represented as shown in Fig. 5.4. A generic discrete-time signal processor (as opposed to a digital signal processor) can be visualized by combining encoder and decoder blocks with the digital computer. This approach allows us to concentrate our focus on the algorithms used in signal processing, instead of the actual digital hardware on which these algorithms are implemented.

5.1 SAMPLING OF CONTINUOUS-TIME SIGNALS

Sampling is the process of obtaining a discrete sequence from an analog signal. Let the analog signal $x_a(t)$ be used as input to an ideal sampler. The output of the sampler would be a discrete sequence $x(n)$ the samples of which are amplitude measurements taken from $x_a(t)$ with increments of T seconds; that is,

$$x(n) = x_a(nT). \quad (5.1)$$

The parameter T is called the *sampling period* or the *sampling interval*. Its reciprocal $f_s = 1/T$ represents the number of samples taken from $x_a(t)$ per unit time interval and is called the *sampling frequency* or the *sampling rate*.

If the sequence $x(n)$ is to be used as an approximation to the analog signal $x_a(t)$, then the sampling process must be reversible, meaning that we should be able to reconstruct the

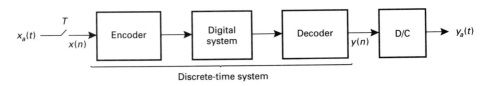

Figure 5.4 Mathematical model for digital processing of analog signals.

Sec. 5.1   Sampling of Continuous-time Signals   85

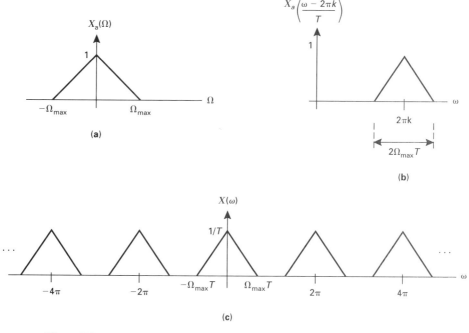

Figure 5.5 Frequency-domain representation of sampling: (a) spectrum of a band-limited analog signal; (b) various terms in Eq. (5.6); (c) spectrum of the sampled signal.

analog signal $x_a(t)$ from the sequence $x(n)$. This requires that the spectral characteristics of the signal be preserved during sampling.

The continuous Fourier transform of the signal $x_a(t)$ is

$$X_a(\Omega) = \int_{-\infty}^{\infty} x_a(t) e^{-j\Omega t} \, dt. \tag{5.2}$$

If the transform $X_a(\Omega)$ is given, the time-domain signal $x_a(t)$ can be found using the inverse Fourier transform

$$x(t) = \frac{1}{2\pi} \int_{-\infty}^{\infty} x_a(t) e^{j\Omega t} \, d\Omega. \tag{5.3}$$

For the discrete sequence $x(n)$, the corresponding transform relationships are

$$X(\omega) = \sum_{n=-\infty}^{\infty} x(n) e^{-j\omega n} \tag{5.4}$$

and

$$x(n) = \frac{1}{2\pi} \int_{-\pi}^{\pi} x(n) e^{j\omega n} \, d\omega. \tag{5.5}$$

Using (5.1) in conjunction with (5.3) and (5.5), it can be shown that the Fourier transforms of $x_a(t)$ and $x(n)$ satisfy the relationship

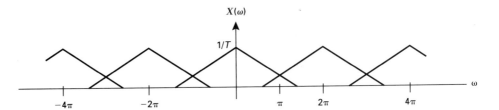

Figure 5.6 Aliasing effect in the frequency domain.

$$X(\omega) = \frac{1}{T} \sum_{k=-\infty}^{\infty} X_a\left(\frac{\omega + 2\pi k}{T}\right). \tag{5.6}$$

Graphical interpretation of (5.6) is important. Figure 5.5 shows the various steps in determining the Fourier transform of a sequence $x(n)$ obtained by sampling a signal $x_a(t)$ with a band-limited triangular-shaped spectrum. Note that the triangular shape of the spectrum is not critical, but the important thing is the fact that the spectrum is band limited; that is, there are no spectral components outside the frequency range $-\Omega_{max} \leq \Omega \leq \Omega_{max}$.

As illustrated in Fig. 5.5, the spectrum of the sampled signal is formed through periodic repetitions of the frequency-scaled spectrum of the analog signal. If the analog signal is to be reconstructed from its samples later, care must be taken to ensure that these periodic repetitions of the analog spectrum do not overlap. If such an overlap occurs, the shape of the analog signal spectrum is not preserved in the spectrum of the sampled signal, and accurate reconstruction is not possible. This effect is known as *aliasing* and is illustrated in Fig. 5.6. To prevent aliasing, the inequality

$$\Omega_{max} T \leq \pi \tag{5.7}$$

must be satisfied. Using $\Omega_{max} = 2\pi f_{max}$, (5.7) can be written in a simpler form:

$$f_s \geq 2f_{max} \tag{5.8}$$

Thus, the sampling rate must be equal to or greater than twice the highest frequency component in the analog signal. The critical rate $2f_{max}$ is known as the *Nyquist rate*. In commercial A/D converters, the analog signal is usually filtered prior to sampling to prevent the aliasing effect.

Exercise 5.1.1

PC-DSP does not have built-in functions that operate on analog signals; however, such functions can be easily developed and incorporated into the environment using the macro capabilities of the program. A number of macros have been included in the distribution disks to aid in the discussion of sampling. These macros are executed by going through the menu selections *Macros/Run-macro* and selecting the desired macro from the list presented.

(a) A continuous-time signal with the name *TestSig* is available. This signal can be tabulated and graphed using the macros *ContTabl* and *ContPlot*, respectively. Additionally, the continuous-time Fourier transform of the signal can be computed using the macro *ContFour*. Compute and graph the Fourier spectrum of this signal. Observe the fact that the signal is almost band limited.

Sec. 5.2   Reconstruction   87

What is the highest significant frequency component? What is the Nyquist sampling rate for this signal?

(b) Sample the signal using a sampling rate that is 25% higher than the Nyquist rate. The macro *SampSig* should be used for this purpose. Compute and graph the DTFT of the discrete-time sequence obtained. Is the spectral shape of the analog spectrum present in the spectrum of the discrete-time signal?

(c) Repeat part (b) using a sampling rate that is 25% lower than the Nyquist rate. How does the DTFT of the signal compare to that obtained in part (b)?

Exercise 5.1.2

Consider the 100-Hz continuous-time sinusoid given by

$$x_a(t) = \sin[2\pi(100)t].$$

This signal is already available on your disk under the name *TestSig2*.

(a) A sequence $x(n)$ is to be obtained by sampling this signal at the rate $f_s = 500$ Hz. What is the normalized frequency of the resulting discrete-time sinusoid?

(b) First, without the aid of the computer, sketch the magnitude and the phase of the DTFT of $x(n)$. Afterward, use PC-DSP to do the same and compare the answers. From the graph of the DTFT, determine the normalized frequency of the discrete-time sinusoid by locating the peak of the spectrum. Note that the magnitude spectrum consists of two lobes, as opposed to impulses. This is due to the fact that, in computer simulation, we are only able to represent a finite-length portion of a sinusoid.

(c) Repeat parts (a) and (b) using the sampling rate $f_s = 140$ Hz. What is the normalized frequency of the discrete-time sinusoid in this case? If an analog signal were to be constructed from the samples of this sequence, what would its frequency be?

5.2 RECONSTRUCTION

Reconstruction is the process of obtaining an analog signal from a discrete-time sequence. In this section, we will mainly concentrate on the second component of a D/A converter, which was termed a discrete-to-continuous converter. Conversion of a discrete-time sequence into an analog signal is accomplished in two steps: First, the discrete-time sequence is converted into a continuous-time impulse train in which impulses are spaced T seconds apart (recall that T is the sampling period), and the area under each impulse is equal to the amplitude of the corresponding sample. Afterward, an analog filter is applied to this train of impulses to obtain a smooth signal. Figure 5.7 illustrates this process. The impulse train obtained from $y(n)$ is

$$s_a(t) = \sum_k y(k)\delta(t - kT). \tag{5.9}$$

Applying the continuous Fourier transform operator to both sides of (5.9) we obtain

$$\begin{aligned} S_a(\Omega) &= \sum_k y(k) \int_t \delta(t - nT) e^{-j\Omega t}\, dt \\ &= \sum_k y(k) e^{-j\Omega nT}. \end{aligned} \tag{5.10}$$

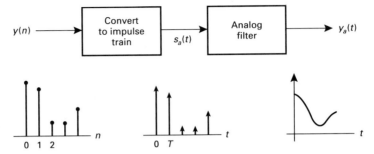

Figure 5.7. Ideal reconstruction process.

The right side of (5.10) is the DTFT of the sequence $y(n)$ evaluated at $\omega = \Omega T$. Thus, the frequency spectra of the sequence $y(n)$ and the impulse train $s_a(t)$ are related by

$$S_a(\Omega) = Y(\Omega T). \tag{5.11}$$

This relationship is illustrated in Fig. 5.8 using the example triangular-shaped spectrum that was also used in Section 5.1. The Fourier transform of the analog output signal $y_a(t)$ is

$$Y_a(\Omega) = S_a(\Omega)G_a(\Omega). \tag{5.12}$$

For $Y_a(\Omega)$ to be identical to $X_a(\Omega)$, the reconstruction filter should be an ideal low-pass filter with a cutoff frequency of $\Omega_c = \pi/T$ rad/s and a passband gain of T.

The impulse response of the reconstruction filter is

$$g_a(t) = \frac{\sin(\pi t/T)}{\pi t/T}. \tag{5.13}$$

The reconstruction signal $y_a(t)$ is the continuous convolution of $s_a(t)$ with this impulse response; that is,

$$y_a(t) = \int_\tau s_a(\tau) g_a(t - \tau) \, d\tau. \tag{5.14}$$

Substituting (5.9) and (5.13) into (5.14) and simplifying, we obtain

$$y_a(t) = T \sum_k y(k) \frac{\sin[\pi(t - kT)/T]}{\pi(t - kT)}. \tag{5.15}$$

Thus, $y_a(t)$ is obtained from $y(n)$ using a $\sin(x)/x$ type of function to interpolate between known amplitudes. This is called *band-limited interpolation*. Unfortunately, the reconstruction filter described by the impulse response in (5.13) is noncausal and therefore physically unrealizable. Relatively good approximations to (5.15) are possible using high-order analog low-pass filters with sharp cutoff characteristics, but this approach is usually expensive and not well suited for mass production. Instead, most commercial D/A converters employ either *zero-order-hold* or *first-order-hold* interpolation. A zero-order-hold reconstruction filter is characterized by the impulse response

$$g_a(t) = u(t) - u(t - T) \tag{5.16}$$

Sec. 5.2   Reconstruction

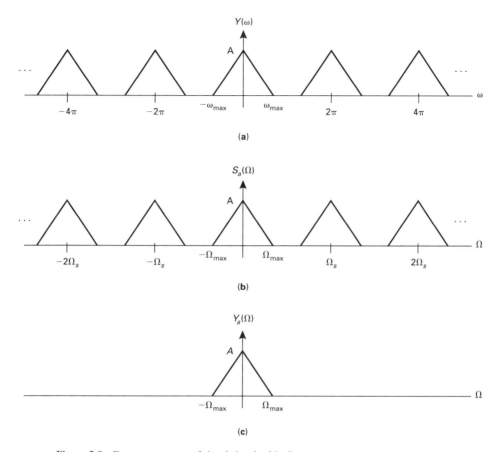

Figure 5.8 Frequency spectra of signals involved in discrete-to-continuous conversion.

and produces an output signal in which the value of each sample is upheld for one sampling period. A first-order-hold reconstruction filter has the impulse response

$$g_a(t) = \begin{cases} t/T + 1, & \text{if } -T \leq t \leq 0 \\ -t/T + 1, & \text{if } 0 \leq t \leq T \\ 0, & \text{otherwise.} \end{cases} \quad (5.17)$$

In this case, the gaps between impulses of $s_a(t)$ are filled by linear interpolation. In some cases, a simple analog low-pass filter is used after zero- or first-order-hold interpolation to smooth the edges. Impulse responses of zero- and first-order-hold reconstruction filters are shown in Fig. 5.9.

Exercise 5.2.1

In this exercise, we will use the two sequences that were obtained in Exercise 5.1.1 by sampling the continuous-time signal *TestSig*.

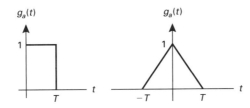

Figure 5.9 Impulse responses of interpolation filters: (a) zero-order-hold; (b) first-order-hold.

(a) First, consider the sequence obtained in part (b) of Exercise 5.1.1. Recall that the sampling rate used was 25% higher than the Nyquist rate. We will construct an analog signal from this sequence. The macro *Reconstr* is provided for this purpose. Most of the parameters required by this macro are self-explanatory. The only parameter that needs explanation is the *step size*. PC-DSP approximates analog signals by means of their samples taken with increments equal to the specified step size T_s. For a good approximation to the analog signal, the step size must be significantly less than the sampling interval T. As a rule of thumb, $T_s \leq 0.1T$ usually works well. Generate an analog signal from the sequence under consideration using zero-order-hold interpolation. Use the same sampling period that was used in Exercise 5.1.1. Graph the resulting analog signal and compare it to the original analog signal that was sampled.

(b) Compute and graph the Fourier transform of the reconstructed signal. Is it similar to the spectrum of the original analog signal?

(c) Repeat parts (a) and (b) using the sequence obtained in part (c) of Exercise 5.1.1. Recall that the sampling rate was 25% lower than the Nyquist rate in this case. How does the reconstructed signal compare to the original? How do the frequency spectra compare? Indicate the area of the spectrum that contains aliasing.

Exercise 5.2.2

Repeat the requirements of Exercise 5.2.1 using first-order-hold instead of zero-order-hold interpolation. How does this affect the appearance of the reconstructed signal when compared to zero-order hold?

Exercise 5.2.3

(a) In part (a) of Exercise 5.1.2, a 100-Hz continuous-time sinusoid was sampled at a rate of $f_s = 500$ Hz. Reconstruct this signal from its samples using zero-order-hold interpolation. Graph the reconstructed signal. Compute and graph its Fourier transform. Are the locations of spectral peaks correct?

(b) In part (c) of Exercise 5.1.2, the same signal was sampled at the rate $f_s = 140$ Hz. Attempt to reconstruct the analog signal from this undersampled sequence. How do the results compare to those obtained in part (a)?

5.3 QUANTIZATION

In signal-processing theory, sample amplitudes of a discrete-time sequence are assumed to have infinite precision; that is, each sample of a sequence can take on a value from a set of infinitely many distinct amplitude levels. On the other hand, digital signal processors

Sec. 5.3 Quantization

are only capable of representing a finite number of distinct amplitude levels. For example, if 4 bits are used to represent the amplitude of each sample of a sequence, then only 16 distinct amplitude levels are possible. In general, a processor that uses b bits for each sample can represent $L = 2^b$ amplitude levels.

The process of converting numbers with infinite precision into numbers from a finite set is called *quantization*. For example, in a system that uses 4 bits per sample, the amplitude range of the input sequence could be divided into 16 equal quantization intervals, and the midpoint of each interval could be taken as a quantization level. In this case, the quantization scheme would consist of representing each sample of the sequence with the quantization level that is closest to it.

If the amplitudes of samples are in the range (x_{min}, x_{max}), and L quantization levels are to be used, then the quantization interval step size is

$$\Delta = \frac{x_{max} - x_{min}}{L} \qquad (5.18)$$

and allowable amplitude levels are

$$q_i = x_{min} + \frac{\Delta}{2} + i\Delta, \qquad i = 0, \ldots, L - 1. \qquad (5.19)$$

Two types of quantizer input–output characteristics are shown in Fig. 5.10. The fundamental difference between the two types is in the way zero-amplitude samples are handled. The first type of quantizer shown is a *midtread* quantizer in which one of the quantization levels has zero amplitude. The second type is a *midriser* quantizer. In this case, there is no zero-amplitude quantization level, and therefore samples with very small amplitudes are quantized to $\mp\Delta/2$ depending on their signs.

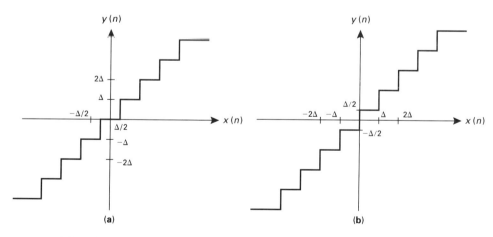

Figure 5.10 Quantizer input–output characteristics: (a) midtread-type quantizer; (b) midriser-type quantizer.

Exercise 5.3.1

(a) Using the formula-entry method of PC-DSP accessed with menu selections *Sequences/Generate-sequence/Formula-entry-method*, create a 200-sample sequence $x(n)$ as

$$x(n) = 10e^{-0.01n} \sin(0.1n), \quad n = 0, \ldots, 199.$$

(b) Quantize samples of this sequence into 16 quantization levels between -10.0 and 10.0. The quantizer function of PC-DSP can be accessed with menu selections *Operations/Nonlinear-operations/Quantize-sequence*. Tabulate and graph the quantized sequence. Identify all quantization levels. Does this correspond to a midriser type or a midtread type of quantizer? Explain.

Since all samples that fall into a quantization interval are assigned the same value (the midpoint of that interval), quantization results in a loss of accuracy. The quantizer output sequence $\tilde{x}(n)$ can be thought of as the sum of the input sequence $x(n)$ and an error signal. The error due to quantization is

$$e_q(n) = \tilde{x}(n) - x(n). \tag{5.20}$$

The amplitude of each sample of $e_q(n)$ is in the range

$$-\frac{\Delta}{2} \leq e_q(n) \leq \frac{\Delta}{2}. \tag{5.21}$$

In most applications that involve quantization, the mean-square value (or power) of the quantization error signal is of interest. This requires knowledge of the probability density function (pdf) of the error signal. Each sample of the error signal takes on a random value in the range given by (5.21). If there is no particular reason to believe that some error values in this range are more likely than others, then each sample of the error signal can be modeled with a random variable uniformly distributed in the range given by (5.21). In this case, the mean (average) value of the quantization error signal is equal to zero. It can be shown that quantization error power is

$$E\{e_q^2(n)\} = \frac{\Delta^2}{12}, \tag{5.22}$$

where $E\{\ldots\}$ denotes the statistical expectation operator. For a finite-length signal, an estimate of the error power in (5.22) can be obtained as

$$P_q = \frac{\sum_{n=0}^{N-1} e_q^2(n)}{N}. \tag{5.23}$$

Exercise 5.3.2

Consider the signal $x(n)$ and its quantized version obtained in Exercise 5.3.1.

(a) Compute the quantization error signal by subtracting the original signal $x(n)$ from its quantized version. Graph the quantization error signal. Does it look "random"?

(b) Obtain an estimate of the error signal power. The mean-square value obtained through menu selections *Operations/Statistics/Signal-statistics* can be used as an estimate of the signal power. Note that this is the approach described by (5.23). How does the value obtained compare to the theoretical value given by (5.22)?

Sec. 5.3   Quantization

The signal-to-noise ratio (SNR) that is due to quantization error is defined as the ratio of the signal power to the noise power; that is,

$$\text{SNR} = \frac{E\{x^2(n)\}}{E\{e_q^2(n)\}}. \tag{5.24}$$

For a finite-length signal $x(n)$, an estimate of the SNR is

$$\text{SNR} \approx \frac{\sum_{n=0}^{N-1} x^2(n)}{\sum_{n=0}^{N-1} e_q^2(n)}. \tag{5.25}$$

Often the signal-to-noise ratio is expressed on a decibel scale as

$$\text{SNR}_{dB} = 10 \log_{10} (\text{SNR}) \tag{5.26}$$

Exercise 5.3.3

Consider again the signal $x(n)$ and its quantized version obtained in Exercise 5.3.1.

(a) Obtain an estimate of the signal power using the same approach that was used for estimating the error power in Exercise 5.3.2. What is the signal-to-noise ratio?

(b) The SNR obtained in part (a) is an average value over 200 samples of the signal. Some signals have instantaneous power levels that vary widely in time. As a result, the SNR due to quantization may also fluctuate in time. Speech and music signals are examples of this phenomenon. As an estimate of the *instantaneous* SNR, we will use

$$\text{SNR}(n) \approx \frac{\sum_{m=0}^{M-1} x^2(n-m)}{\sum_{m=1}^{M-1} e_q^2(n-m)}, \tag{5.27}$$

where $M < N$. Technically, this is a rather crude estimate of (5.24) for nonstationary signals. Using PC-DSP, compute and plot this estimate with $M = 20$. First, square the signals $x(n)$ and $e_q(n)$ using the multiplication function in the *Operations/Arithmetic-operations* menu. Afterward, generate a sequence of 20 unit-amplitude samples, and convolve it with the squared signals to obtain the summations in the preceding SNR(n) expression. Finally, obtain the SNR estimate using the division function in the *Operations/Arithmetic-operations* menu. It may be necessary to add a very small positive offset to the denominator sequence to avoid division by zero. Comment on the variation of the SNR estimate. Is it what you expected?

As the preceding exercise demonstrates, the signal-to-noise ratio due to quantization is approximately proportional to the signal power and is therefore degraded in weaker portions of the signal being quantized. This is due to the fact that the quantization error power remains fairly constant throughout, independent of the signal power. One possible solution to this problem involves choosing quantization levels nonuniformly so that lower-amplitude samples are quantized more finely than higher-amplitude samples. Alternatively, conditioning the signal with a nonlinear characteristic and then using a uniform quantizer achieves the same result. One such characteristic is the μ-law compression formula

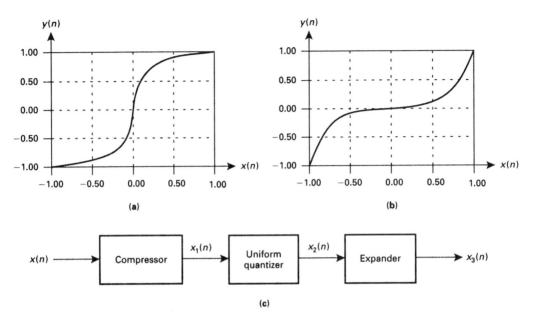

Figure 5.11 Nonuniform quantization: (a) μ-law compressor characteristic; (b) μ-law expander characteristic; (c) use of μ-law characteristics to achieve nonuniform quantization.

$$y(n) = \frac{\text{sgn}(x)}{\ln(1+\mu)} \ln\left(1 + \mu\frac{|x|}{x_{\max}}\right), \quad |x| \leq x_{\max}, \qquad (5.28)$$

which compresses the dynamic range of the signal $x(n)$ by amplifying low-amplitude samples more heavily than higher-amplitude samples. The new signal $y(n)$ can now be uniformly quantized, yielding a more constant SNR when compared to $x(n)$. The dynamic range of the signal $y(n)$ can be expanded using the inverse of (5.28):

$$x = \frac{x_{\max}\text{sgn}(y)}{\mu}[(1+\mu)^{|y|} - 1] \qquad (5.29)$$

The use of nonuniform quantization is shown in Fig. 5.11 along with μ-law compressor and expander characteristics.

Exercise 5.3.4

(a) In this exercise, we will repeat the requirements of Exercise 5.3.3 using nonuniform quantization. To achieve this, we will first apply the μ-law compression formula given by (5.28) to the signal $x(n)$. In PC-DSP, use menu selections *Operations/Nonlinear-operations/Compander* and select the μ-law compressor option with the parameter value $\mu = 100$ to obtain the signal $x_1(n)$. Now quantize this compressed signal into 16 amplitude levels as in the previous exercise, thus obtaining the signal $x_2(n)$. In the final step, apply the μ-law expansion formula given by (5.29) to $x_2(n)$ to obtain the nonuniformly quantized signal $x_3(n)$. For this step, use the menu selections *Operations/Nonlinear-operations/Compander* and select the μ-law expander option

Sec. 5.4   Changing the Sampling Rate   95

with the parameter value $\mu = 100$. Tabulate and graph the resulting sequence. Identify the quantization levels.

(b) Obtain the quantization error signal by subtracting the original signal $x(n)$ from the quantized signal $x_3(n)$. Graph the error signal. Do the sample amplitudes of the error signal appear to be uniformly distributed?

(c) Obtain an estimate of the instantaneous signal-to-noise ratio using the technique outlined in Exercise 5.3.3. How does the SNR estimate differ from that obtained in the previous exercise? Comment.

Exercise 5.3.5

In this exercise, we will observe μ-law compressor characteristics for various values of the parameter μ. The easiest method of doing this is to generate a sequence $x(n)$ as a ramp sequence with unit slope, to apply μ-law compression to it, and to graph the resulting sequence. Generate the 100-sample sequence

$$x(n) = n\, u(n), \qquad n = 0, \ldots, 99.$$

Apply μ-law compression to $x(n)$ for parameter values $\mu = 0, 20, 50, 100, 150$, and 200. Graph the output sequence for each case. Comment on how the value of μ affects the shape of the characteristic.

5.4 CHANGING THE SAMPLING RATE

In some digital signal-processing applications, it may be necessary to increase or decrease the sampling rate of a discrete-time signal. In large-scale systems, various components of the system might process signals at different sampling rates, requiring sampling-rate conversions to be performed. The sampling rate of a discrete-time signal may be altered using one of two different methods. In the first method, the discrete-time signal is converted to a continuous-time signal through the use of a D/A converter, and this continuous-time signal is resampled at the desired new rate using an A/D converter. This method is expensive to implement since it requires a D/A and A/D converter pair for each alteration of the sampling rate. Also, the quality of the signal is degraded somewhat due to the distortion introduced in A/D and D/A conversion operations. The second method of altering the sampling rate is to process the discrete-time signal without converting it to analog. Usually, this is the preferred method of changing the sampling rate on a discrete-time signal.

A system that reduces the sampling rate of a discrete-time signal by an integer ratio is called a *decimator*. Conversely, a system that increases the sampling rate of a discrete-time signal by an integer ratio is called an *interpolator*. Two fundamental building blocks used in practical decimators and interpolators are *downsamplers* and *upsamplers* (Fig. 5.12).

A *downsampler* has the input–output relationship

$$x_D(n) = x(nD), \qquad (5.30)$$

where the downsampling rate D is an integer. Out of every D samples of the input signal $x(n)$, only one appears at the output of the downsampler, and the remaining $D - 1$

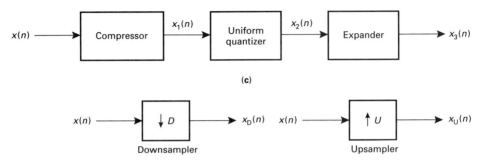

Figure 5.12 Building blocks used in sampling rate alteration.

samples are discarded. Thus, the sampling rate of the output signal is D times lower than that of the input signal. Here, the important question that needs to be addressed is whether or not the discarded samples are replaceable. In other words, are we losing information by discarding some of the input samples, or is this merely removal of redundancy? The answer can be found by studying the relationship between the frequency spectra of $x(n)$ and $x_D(n)$. It can be shown that

$$X_D(\omega) = \frac{1}{D}\sum_{k=0}^{D-1} X\left(\frac{\omega + 2\pi k}{D}\right). \qquad (5.31)$$

This relationship is depicted graphically in Fig. 5.13. Note the similarity of (5.31) to (5.6) given in Section 5.1. Clearly, if the downsampling rate D is too large, that is, if too many samples are discarded, the frequency spectrum of the output signal represents an aliased version of the spectrum of the input signal. In this case, the original signal $x(n)$ cannot be reconstructed from its downsampled version $x_D(n)$.

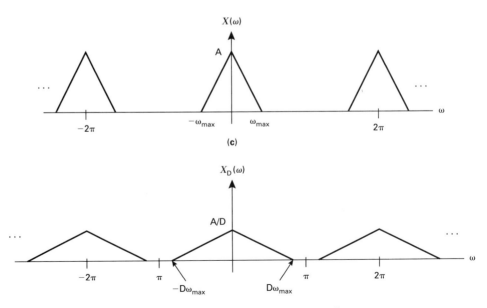

Figure 5.13 Spectral relationships in downsampling.

Sec. 5.4 Changing the Sampling Rate

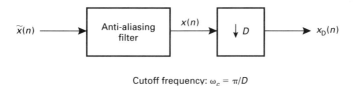

Cutoff frequency: $\omega_c = \pi/D$

Figure 5.14 Decimator for reducing the sampling rate by a factor of D.

To prevent aliasing, the input signal $x(n)$ must be band limited such that

$$X(\omega) = 0, \quad \text{for } |\omega| \leq \pi/D. \tag{5.32}$$

In a practical decimator, the input signal is low-pass-filtered prior to the downsampling operation to ensure that the band-limiting condition given by (5.32) is satisfied. Thus, a decimator consists of an antialiasing filter and a downsampler connected in cascade (Fig. 5.14).

Exercise 5.4.1

(a) A discrete sequence with the name *Ex-5-4-1* is stored in the data file EX5-4-1.SEQ on your program disks. Using PC-DSP, graph this sequence. Also compute and graph its discrete-time Fourier transform. Looking at the magnitude of the spectrum, determine the largest downsampling rate that can be used on this sequence without causing aliasing.

(b) Looking at the magnitude and the phase of the spectrum in part (a), roughly sketch what the spectrum would look like if this sequence were downsampled with $D = 2$. Repeat with $D = 4$.

(c) Downsample the sequence with $D = 2$ using menu selections *Operations/Processing-functions/Downsample-sequence*. Graph the downsampled sequence and compare it to the original. Compute and graph its DTFT and use it to check your hand sketch in part (b). Does there seem to be aliasing in the output spectrum?

(d) Repeat part (c) with $D = 4$. Is aliasing present now?

Exercise 5.4.2

The downsampling operation in part (d) of Exercise 5.4.1 results in an aliased output spectrum. As previously discussed, aliasing can be prevented by band limiting the input sequence prior to downsampling. A finite-impulse-response low-pass filter with a cutoff frequency $\omega_c = \pi/4$ is stored in the data file ANTIALIA.FLT.

(a) Process the input signal *Ex5-4-1* with the antialiasing filter. Use menu selections *Filters/Simulate-filter*. In the data-entry window presented, specify the names of the input signal and the filter. Graph the band-limited signal.

(b) Compute and graph the frequency spectrum of the band-limited signal. Looking at the spectrum, roughly sketch what the spectrum would look like if this band-limited sequence were downsampled with $D = 4$.

(c) Downsample the band-limited signal with $D = 4$. Graph the downsampled signal and compare to the band-limited signal obtained in part (a).

(d) Compute and graph the frequency spectrum of the band-limited signal. Check this against your hand sketch in part (b). Is there aliasing in the output spectrum?

An *upsampler* is defined by the input–output relationship

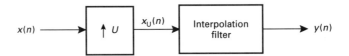

Figure 5.15 Interpolator for increasing the sampling rate by a factor of U.

$$x_U(n) = \begin{cases} x(n/U), & n = kU \\ 0, & \text{otherwise}, \end{cases} \quad (5.33)$$

where U is an integer upsampling rate. The output signal is obtained by inserting zero-amplitude samples between the existing samples of the input signal. A total of $U - 1$ zero-amplitude samples are inserted between any two consecutive samples of $x(n)$. Thus, the sampling rate of the output signal $x_U(n)$ is U times higher than that of the input signal $x(n)$. However, $x_U(n)$ does not look similar to $x(n)$. It needs to be processed through an interpolation filter so that the zero-amplitude samples inserted in the upsampling process are modified to "blend in" with the rest of the signal. An interpolator for increasing the sampling rate consists of an upsampler and an interpolation filter connected in cascade (Fig. 5.15).

The frequency spectrum of the upsampled signal is

$$X_U(\omega) = X(\omega U).$$

Figure 5.16 illustrates the relationship between the spectra of the upsampler input and output signals. Clearly, the optimum choice for the interpolation filter is an ideal low-pass filter with cutoff frequency $\omega_c = \pi/U$. In practice, a filter that approximates this ideal characteristic is usually adequate.

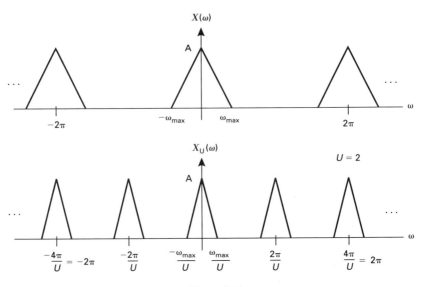

Figure 5.16

Sec. 5.4 Changing the Sampling Rate

Exercise 5.4.3

(a) A discrete sequence with the name *Ex-5-4-3* is stored in the data file EX5-4-3.SEQ on your program disks. Using PC-DSP, graph this sequence. Also compute and graph its frequency spectrum (DTFT).

(b) Looking at the magnitude and the phase of the spectrum, roughly sketch what the spectrum would look like if this sequence were upsampled with $U = 2$. Repeat with $U = 3$.

(c) Upsample the sequence with $U = 2$ using menu selections *Operations/Processing-functions/ Upsample-sequence*. Name the resulting signal *Upsamp2*. Graph the upsampled sequence and compare it to the original. Compute and graph its DTFT and use it to check your hand sketch in part (b).

(d) Repeat part (c) with $U = 3$ and name the resulting signal *Upsamp3*.

Exercise 5.4.4

In this exercise, we will process the upsampled signals of Exercise 5.4.3 with appropriate interpolation filters so that the zero-amplitude samples inserted in the upsampling process are modified.

(a) Consider an FIR filter with the impulse response

$$h(n) = \delta(n) + \delta(n - 1).$$

This is a zero-order-hold interpolation filter. Using PC-DSP, generate the two-sample impulse response of this filter. Compute the frequency response of the filter using the DTFT function. Graph the magnitude and the phase of the frequency response. At which frequency does the first zero crossing of the magnitude occur? You should find out that this filter is a crude approximation to an ideal interpolation filter for a signal upsampled with $U = 2$.

(b) Process the signal *Upsamp2* obtained in part (c) of Exercise 5.4.3 with the zero-order-hold interpolation filter using the convolution function. Recall that the convolution function is accessed with menu selections *Operations/Processing-functions/Convolve-sequences*. Graph the resulting sequence. Tabulate the original and filtered sequences side by side and compare the sample values.

(c) In part (d) of Exercise 5.4.3, the signal *Upsamp3* was obtained by upsampling the original signal with $U = 3$. A zero-order-hold interpolation filter for this case would have the impulse response

$$h(n) = \delta(n) + \delta(n - 1) + \delta(n - 2).$$

Repeat parts (a) and (b) using the signal *Upsamp3* with this three-tap interpolation filter.

Exercise 5.4.5

In the previous exercise, zero-order-hold interpolation was used to modify the zero-amplitude samples inserted into the signal in the upsampling process. In this exercise, we will use a slightly better approximation to the ideal interpolation filter.

(a) Consider an FIR filter with the impulse response

$$h(n) = 0.5\delta(n + 1) + \delta(n) + 0.5\delta(n - 1).$$

This is a first-order-hold interpolation filter. Using PC-DSP, generate the three-sample impulse response of this filter. Compute and graph the frequency response of the filter. At which frequency does the first zero crossing of the magnitude occur? You should find out that this filter is a somewhat better approximation to an ideal interpolation filter for a signal upsampled with $U = 2$.

(b) Process the signal *Upsamp2* obtained in part (c) of Exercise 5.4.3 with the first-order-hold interpolation filter using the convolution function. Graph the resulting sequence. Tabulate the original and filtered sequences side by side and compare the sample values. Note how the zero-amplitude samples are interpolated linearly between existing samples.

(c) A similar first-order-hold interpolation filter can be obtained for processing the signal *Upsamp3*. In this case, the impulse response is

$$h(n) = \frac{1}{3}\delta(n+2) + \frac{2}{3}\delta(n+1) + \delta(n) + \frac{2}{3}\delta(n-1) + \frac{1}{3}\delta(n-2).$$

Repeat parts (a) and (b) using the signal *Upsamp3* with this filter.

6

The z-Transform and Discrete-time Structures

The important thing is not to stop questioning.
Curiosity has its own reason for existing.
—Albert Einstein

The z-transform can be viewed as the discrete-time counterpart of the Laplace transform that is used for continuous-time signals. We will see that, as a transform, the z-transform is more general than the discrete-time Fourier transform (DTFT). The latter can be thought of as a special case of the former. Also, in analysis and design problems, we will find the z-transform version of the system function concept to be somewhat more convenient to use compared to the DTFT-based system function concept that was discussed in Chapter 3. We will also discuss real-time implementation methods for discrete-time systems described by means of a z-domain system function.

6.1 BASIC CONCEPTS

The z-transform of a sequence $x(n)$ is defined as

$$X(z) = \sum_{n=-\infty}^{\infty} x(n) z^{-n}, \qquad (6.1)$$

where z is a complex variable; that is, $z = re^{j\omega}$. One method of geometrically representing the complex variable z is to associate it with a point on the complex plane such that r is

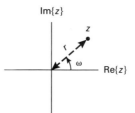

Figure 6.1 The complex plane.

the distance of this point from the origin, and ω is the angle with the horizontal axis, measured counterclockwise. Figure 6.1 illustrates this representation.

Using the polar form of the complex variable z, the transform in (6.1) can be written as

$$X(r,\omega) = \sum_{n=-\infty}^{\infty} x(n)(re^{j\omega})^{-n}$$

$$= \sum_{n=-\infty}^{\infty} \underbrace{[x(n)r^{-n}]}_{\tilde{x}(n)} e^{-j\omega n}$$
(6.2)

Equation (6.2) provides an interesting way to look at the z-transform. It demonstrates that the z-transform of a sequence $x(n)$ evaluated at the point $z = re^{j\omega}$ is equal to the discrete-time Fourier transform (DTFT) of the modified sequence $\tilde{x}(n) = x(n)r^{-n}$ evaluated for the angular frequency value ω. If the parameter r is chosen to be equal to unity, then (6.2) simply reduces to the DTFT of $x(n)$. Furthermore, if the parameter ω is varied between 0 and 2π radians, corresponding values of the complex variable $z = re^{j\omega}$ lie on a circle in the complex plane, with its center at the origin and with a radius equal to unity. Hence, the relationship between the DTFT and the z-transform is clear. The DTFT of a sequence is equal to its z-transform evaluated for values of z that are on the unit circle of the complex z-plane.

The z-transform defined by (6.1) is an infinite series, and divergence problems may occur for some values of the complex variable z. We might intuitively guess from (6.2) that the convergence of the transform should depend on the value of the parameter r and not on the value of ω. In general, the *region of convergence* for the transform is in the form

$$R_1 < |z| < R_2,$$

where $|z| = r$. The boundaries R_1 and R_2 depend on the signal being transformed. The special cases where $R_1 \to 0$ and/or $R_2 \to \infty$ can also be represented using this general form. The significance of the region of convergence concept will be illustrated with an example.

Consider the right-sided exponential sequence

$$x(n) = a^n u(n).$$

Using (6.1), the z-transform of $x(n)$ is

$$X(z) = \sum_{n=-\infty}^{\infty} a^n u(n) z^{-n}.$$

Sec. 6.1   Basic Concepts   103

Dropping the factor $u(n)$ and adjusting the summation limits, we obtain

$$X(z) = \sum_{n=0}^{\infty} a^n z^{-n},$$

which can be put in closed form as

$$X(z) = \frac{1}{1 - az^{-1}} \qquad (6.3)$$

subject to the constraint $|az^{-1}| < 1$. Thus, the region of convergence is

$$\text{ROC: } |z| > |a|.$$

On the z-plane, this inequality represents all points that are *outside* the circle centered around the origin with radius equal to $|a|$.

As another specific example, consider the left-sided exponential sequence

$$w(n) = -a^n u(-n - 1).$$

Using (6.1), the z-transform of $w(n)$ is

$$W(z) = -\sum_{n=-\infty}^{\infty} a^n u(-n-1) z^{-n}$$

$$= -\sum_{n=-\infty}^{-1} a^n z^{-n}$$

$$= -\sum_{\tilde{n}=1}^{\infty} a^{-\tilde{n}} z^{\tilde{n}}.$$

In the last step, a variable change was made by $\tilde{n} = -n$ so that positive summation limits can be obtained. It can be easily shown that the closed form of $W(z)$ is

$$W(z) = \frac{1}{1 - az^{-1}} \qquad (6.4)$$

subject to the constraint $|a^{-1}z| < 1$. This yields the region of convergence

$$\text{ROC: } |z| < |a|,$$

which corresponds to the *inside* of the circle with radius $|a|$ on the z-plane.

The important thing to note in these examples is that the analytical expressions for $X(z)$ and $W(z)$ are identical. The only difference between the two transforms is in the regions of convergence. Thus, the region of convergence is an integral part of the z-transform and must be specified. If only a transform expression $X(z)$ is given without the region of convergence, there will be an uncertainty about the sequence that led to this transform since there may be more than one possible answer.

For a rational z-transform $X(z)$, the region of convergence is a connected region in the complex plane bounded by the poles of the transform. The general shape of this region depends on the type of the signal.

Right-sided sequence.   A right-sided sequence $x(n)$ satisfies the condition

$$x(n) = 0, \qquad \text{for } n < N_1.$$

For this type of a sequence, the z-transform can be written as

$$X(z) = \sum_{n=N_1}^{\infty} x(n) z^{-n},$$

and the region of convergence is the region of the z-plane that is *outside* a circle, that is,

$$|z| > R_1,$$

with the radius R_1 determined by the pole of the transform $X(z)$ that is farthest from the origin of the z-plane. Additionally, values of z with infinite magnitude have to be excluded from the region of convergence if $N_1 < 0$.

Left-sided sequence. A left-sided sequence $x(n)$ satisfies the condition

$$x(n) = 0, \quad \text{for } n > N_2.$$

For this type of a sequence, the z-transform can be written as

$$X(z) = \sum_{n=-\infty}^{N_2} x(n) z^{-n},$$

and the region of convergence is the region of the z-plane that is *inside* a circle, that is,

$$|z| < R_2,$$

with the radius R_2 determined by the pole of the transform $X(z)$ that is closest to the origin of the z-plane. Additionally, the origin of the z-plane have to be excluded from the region of convergence if $N_2 > 0$.

Two-sided sequence. A two-sided sequence $x(n)$ is one that exists for all values of the index n. It can be written as the sum of a left-sided sequence and a right-sided sequence:

$$x(n) = x_L(n) + x_R(n).$$

The transform $X(z)$ is the sum of the corresponding transforms:

$$X(z) = X_R(z) + X_L(z).$$

The region of convergence is in the form

$$R_1 < |z| < R_2$$

with R_1 determined by the pole of the transform $X_R(z)$ that is farthest from the origin of the z-plane and R_2 determined by the pole of the transform $X_L(z)$ that is closest to the origin of the z-plane. Thus, the region of convergence is the overlap of the two regions, if an overlap exists. Note that the overlap does not exist if $R_1 \geq R_2$.

Finite-length sequence. A finite-length sequence is one that satisfies

$$x(n) = 0, \quad \text{for } n < N_1 \text{ or } n > N_2.$$

The z-transform equation can be written in the form

Sec. 6.1 Basic Concepts

TABLE 6.1 Some Properties of the z-Transform

Signal		Transform
1. $x(n)$	$\Longleftrightarrow$	$X(z)$
2. $x^*(n)$	$\Longleftrightarrow$	$X^*(z^*)$
3. Linearity		
$\alpha x_1(n) + \beta x_2(n)$	$\Longleftrightarrow$	$\alpha X_1(z) + \beta X_2(z)$, all α and β
4. Time shifting		
$x(n - M)$	$\Longleftrightarrow$	$z^{-M}X(z)$
5. Time reversal		
$x(-n)$	$\Longleftrightarrow$	$X(z^{-1})$
6. Convolution		
$x(n)*h(n)$	$\Longleftrightarrow$	$X(z)H(z)$
7. Differentiation in frequency		
$nx(n)$	$\Longleftrightarrow$	$-z\dfrac{d}{dz}[X(z)]$

$$X(z) = \sum_{n=N_1}^{N_2} x(n)z^{-n}.$$

The region of convergence is the entire z-plane, with two possible exceptions: If $N_1 < 0$, values of z with infinite magnitude are excluded. If $N_2 > 0$, the origin of the z-plane is excluded.

Some of the important properties of the z-transform are summarized in Table 6.1. Proofs of these properties will be left to the reader. All properties can be easily proved starting with the definition of the z-transform given by (6.1).

Exercise 6.1.1

Without using computer aid, find the z-transform of each of the following sequences. In each case, indicate the region of convergence (both analytically and graphically).

(a) $x(n) = u(n) - u(n - 7)$
(b) $x(n) = u(n + 1) - u(n - 6)$
(c) $x(n) = (\frac{1}{2})^n u(n) + (\frac{1}{3})^n u(n)$
(d) $x(n) = (\frac{1}{2})^n u(n) - (\frac{1}{3})^n u(-n - 1)$
(e) $x(n) = \sin(\omega_0 n)u(n)$
(f) $x(n) = \cos(\omega_0 n)u(n)$

Hint: For parts (e) and (f), express the signal with complex exponentials using Euler's formula, find the z-transform of each term, and add the results.

We established in Chapter 3 that a DTLTI system can be uniquely described by means of a linear constant-coefficient difference equation in the general form

$$y(n) = -\sum_{k=1}^{N} a_k y(n - k) + \sum_{r=0}^{M} b_r x(n - r). \tag{6.5}$$

By taking the z-transform of (6.5) and rearranging terms, we obtain

$$Y(z)\left[1 + \sum_{k=1}^{N} a_k z^{-k}\right] = X(z) \sum_{r=0}^{M} b_r z^{-r} \qquad (6.6)$$

where linearity and time-shifting properties of the z-transform have been used. The z-domain system function can be computed from (6.6) as

$$H(z) = \frac{Y(z)}{X(z)} = \frac{\sum_{r=0}^{M} b_r z^{-r}}{\sum_{k=1}^{N} a_k z^{-k}}. \qquad (6.7)$$

Exercise 6.1.2

A causal discrete-time system is described with the difference equation

$$y(n) = 0.4y(n-1) + 0.21y(n-2) + x(n).$$

(a) Use the difference equation solution function of PC-DSP to find the impulse response of this system for $0 \leq n \leq 199$. Recall that this function is accessed with menu selections *Operations/Processing-functions/Difference-equation*.

(b) Find the z-domain transfer function $H(z)$ for the system under consideration. Write $H(z)$ using only positive powers of z. Computer use is not necessary for this part.

(c) Enter a description of this system into PC-DSP using menu selections *Filters/Enter-external-filter*. This function creates a discrete-time system description in the proper format so that it can be analyzed using analysis functions of PC-DSP. When the data-entry form is displayed, enter numerator and denominator coefficients in the order of descending powers of z, separated by commas. Also type a name for the discrete-time system entered.

(d) We are now ready to look at the poles and the zeros of the transfer function. Use menu selections *Filters/Analyze-filter*. Type the name of the system entered into PC-DSP in part (c), select *Poles & zeros*, and press the button labeled *Plot*. This produces a pole–zero plot in the z-domain. Alternatively, it is possible to tabulate the poles and the zeros of the system with the button labeled *Tabulate*.

(e) How does the pole–zero plot relate to the impulse response of the system obtained in part (a)? Comment.

Exercise 6.1.3

Repeat the requirements of Exercise 6.1.2 for systems with the difference equations given next. Assume that all systems are causal. In each case, pay special attention to the relationship between the impulse response of the system and the locations of its poles on the z-plane.

(a) $y(n) = 0.9y(n-1) + 0.36y(n-2) + x(n)$
(b) $y(n) = 0.8y(n-1) - 0.25y(n-2) + x(n)$
(c) $y(n) = 1.4y(n-1) - 1.13y(n-2) + x(n)$

Exercise 6.1.4

In some applications, it may be necessary to implement a discrete-time system to produce samples of a sinusoidal signal at a specified frequency. This is known as *frequency synthesis*. A common

Sec. 6.1 Basic Concepts

example is a digital keyboard that produces sounds similar to those of certain musical instruments. For each note, an appropriate sinusoidal signal is generated and shaped to produce the desired sound. In this exercise, we will look at one possible method of generating discrete-time sinusoids in real time.

(a) To generate a discrete-time sinusoid at the frequency ω_0, we will use a system the impulse response of which is

$$h(n) = \sin(\omega_0 n)u(n).$$

Find the z-domain system function $H(z)$ as the z-transform of the signal $h(n)$. Note that this is identical to part (e) of Exercise 6.1.1, so the result may already be available.

(b) Write the system function $H(z)$ using negative powers of z. By inspection of the system function, write a difference equation between the input signal $x(n)$ and the output signal $y(n)$.

(c) Use the difference equation solution function of PC-DSP to compute the impulse response of the system for $0 \leq n \leq 199$ with $\omega_0 = 0.1\pi$ radians. Graph the impulse response obtained. How many samples are in each cycle of the sinusoid?

(d) Repeat part (c) with $\omega_0 = 0.2\pi$ radians.

Exercise 6.1.5

Consider a DTLTI system with a first-order system function in the form

$$H(z) = \frac{z - (1/a^*)}{z - a}.$$

It can be shown that this system has an all-pass magnitude characteristic; that is, the magnitude of the system function is constant independent of the angular frequency ω. The system has a pole and a zero that are mirror images of each other with respect to the unit circle of the z-plane. Higher-order all-pass systems can be constructed by cascading several first-order all-pass systems.

(a) Write the system function for a first-order all-pass system with $a = 0.75$. Enter the resulting system into PC-DSP using menu selections *Filters/Enter-external-filter*. Tabulate and graph the magnitude response of the system. Also observe the phase behavior.

(b) A test signal has been stored in the data file EX6-1-5.SEQ. Graph the signal in the time domain. Afterward, compute and graph the DTFT of the signal.

(c) Process the test signal with the all-pass filter obtained in part (a). This can be done in two different ways: Use either the difference equation solution function accessed with menu selections *Operations/Processing functions/Difference-equation* or the discrete-time filter simulation function accessed with menu selections *Filters/Simulate-filter*. Graph the output signal in the time domain. Is it identical to the input signal? If not, why not?

(d) Compute the DTFT of the output signal and compare it to the DTFT of the input signal. Are magnitude responses the same?

Causality. A causal system is one the impulse response of which satisfies

$$h(n) = 0, \quad \text{for } n < 0.$$

This represents a right-sided sequence that starts at index $N_1 = 0$. As a result, the region of convergence for the z-transform of $h(n)$ must include infinite values of z.

Stability. Recall from Chapter 3 that, for a discrete-time linear and time-invariant system to be stable, its impulse response must be absolutely summable; that is,

$$\sum_{k=-\infty}^{\infty} |h(k)| < \infty.$$

This requires that the magnitude of the z-transform of $h(n)$ be finite for $|z| = 1$. Thus, a discrete-time system is stable if the region of convergence of its z-domain system function includes the unit circle of the z-plane.

Next, we will establish the conditions for a *causal system* to be stable. If the system under consideration is causal, the region of convergence for the system function is the region outside a circle:

$$\text{ROC:} \quad |z| > R_1.$$

If the system is also stable, then the unit circle of the z-plane is inside the region of convergence, and thus $R_1 < 1$. Since there may not be any poles inside the region of convergence, all the poles of the system function must be on or inside the circle with radius R_1. In other words, all poles must be inside the unit circle.

Exercise 6.1.6

Consider a DTLTI system with the system function

$$H(z) = \frac{z(z-1)}{z^2 - 1.4z + 1.13} = \frac{z(z-1)}{(z - 0.7 - j0.8)(z - 0.7 + j0.8)}$$

(a) Enter this system into PC-DSP using menu selections *Filters/Enter-external-filter*. Afterwards, obtain the magnitude and the phase characteristics of the system using menu selections *Filters/Analyze-filter*. Graph the results.

(b) Write a difference equation between the input and the output signals of this system. Use the difference equation solution function of PC-DSP to compute the unit-step response of this system for $0 \leq n \leq 199$. Tabulate and graph the impulse response obtained. Does the system appear to be stable?

(c) In some applications it may be necessary to stabilize an unstable system function without affecting its magnitude response. Devise a second-order all-pass system such that, when it is cascaded with $H(z)$, the resulting system function is stable. *Hint*: Select the zeros of the all-pass system to coincide with the unstable poles of $H(z)$.

(d) Write the system function for the cascade combination of $H(z)$ and the all-pass filter. Simplify it by canceling common factors. Enter it into PC-DSP as was done in part (a). Graph the magnitude and phase characteristics of the stable system. Compare the magnitude response to that obtained in part (a).

(e) Compute the unit-step response of the stable system and compare to that obtained in part (b).

6.2 INVERSE z-TRANSFORM

In this section, we will discuss the problem of finding a sequence $x(n)$ given its z-transform $X(z)$. If the solution is to be unique, the region of convergence must be specified or indirectly implied. We will focus our attention on the case in which the transform is a rational function.

Sec. 6.2 Inverse z-Transform

One method of computing the inverse z-transform is the partial fraction expansion. The transform is written in the form

$$X(z) = \frac{k_1 z}{z - \alpha_1} + \frac{k_2 z}{z - \alpha_2} + \cdots + \frac{k_M z}{z - \alpha_M}, \tag{6.8}$$

and the sequence $x(n)$ is determined by summing inverse transforms of the terms in this expansion. The type of each term (right sided or left sided) is determined based on the location of its pole in relation to the region of convergence of $X(z)$. For a right-sided term, the transform pair

$$\frac{k_i z}{z - \alpha_i} \iff k_i \alpha_i^n u(n) \tag{6.9}$$

is used. For a left-sided term, on the other hand, the transform pair to use is

$$\frac{k_i z}{z - \alpha_i} \iff -k_i \alpha_i^n u(-n - 1). \tag{6.10}$$

Note that, for a transform to be expanded into partial fractions as in (6.8), the order of the numerator polynomial must be less than the order of the denominator polynomial. If this is not the case, the transform $X(z)$ can be written as

$$X(z) = \tilde{X}(z) + Q(z)$$

where $Q(z)$ is a polynomial of z, and $\tilde{X}(z)$ satisfies the condition mentioned previously.

Exercise 6.2.1

(a) A sequence $x(n)$ has the z-transform

$$X(z) = \frac{z - 1}{(z - 0.7)(z - 1.2)}$$

and the region of convergence is $|z| > 1.2$. Using partial fraction expansion, find an analytical expression for $x(n)$.

(b) Repeat part (a) for the case for which the region of convergence is $|z| < 0.7$.

(c) Repeat part (a) for the case for which $x(n)$ is the impulse response of a *stable* system. Note that the region of convergence is implied in this case.

An alternative method of finding the inverse z-transform is long division. This method does not result in an analytical expression for the sequence $x(n)$. It is a numerical method that can be used to compute $x(n)$ one sample at a time. It will be demonstrated using the transform of Exercise 6.2.1. We first need to write $X(z)$ as a ratio of two polynomials; that is,

$$X(z) = \frac{z - 1}{z^2 - 1.9z + 0.84}.$$

The numerator of $X(z)$ can be divided by its denominator iteratively to obtain an infinite-series form of $X(z)$.

$$\begin{array}{r} z^{-1} + 0.9z^{-2} + 0.87z^{-3} \\ z^2 - 1.9z + 0.84 \overline{)z - 1 } \\ \underline{z - 1.9 + 0.84z^{-1}} \\ 0.9 - 0.84z^{-1} \\ \underline{0.9 - 1.71z^{-1} + 0.756z^{-2}} \\ 0.87z^{-1} - 0.756z^{-2} \\ \underline{0.87z^{-1} - 1.653z^{-2} + 0.7308z^{-3}} \\ 0.897z^{-2} - 0.7308z^{-3} \end{array}$$

Thus, after three iterations, we can write $X(z)$ as

$$X(z) = z^{-1} + 0.9z^{-2} + 0.87z^{-3} + \frac{0.897z^{-2} - 0.7308z^{-3}}{z^2 - 1.9z + 0.84}.$$

It is obvious that

$$x(0) = 0, \quad x(1) = 1, \quad x(2) = 0.9, \quad x(3) = 0.87.$$

More samples of $x(n)$ can be obtained by continuing with the long division.

Exercise 6.2.2

(a) Continue the long division already started and obtain samples of $x(n)$ for the range $4 \leq n \leq 10$.

(b) PC-DSP has a built-in function for performing long division. It is accessed with menu selections *Transforms/Inverse-z-transform*. The data-entry form presented has fields for numerator and denominator coefficients. The transfer function we have used can be entered in the following form:

Numerator: 1 ; -1
Denominator: 1 ; -1.9 ; 0.84

Note that numerator and denominator coefficients are ordered in terms of descending powers of z and are separated from each other with semicolons. Select *right-sided sequence*. In addition, initial and final values of the sample index need to be specified. Enter 0 and 10, respectively. Compare the result to your hand calculations in part (a).

In the long-division example, we obtained a right-sided solution for $x(n)$. Long-division technique can also be used for obtaining a left-sided solution if that is what the region of convergence dictates. The trick is to write numerator and denominator polynomials in the order of ascending powers of z.

$$\begin{array}{r} -1.1905 - 1.5023z - 1.9807z^2 \\ 0.84 - 1.9z + z^2 \overline{)-1 + z } \\ \underline{-1 + 2.2619z - 1.1905z^2} \\ -1.2619z + 1.1905z^2 \\ \underline{-1.2619z + 2.8543z^2 - 1.5023z^3} \\ -1.6638z^2 + 1.5023z^3 \end{array}$$

We can conclude that

$$x(0) = -1.1905, \quad x(-1) = -1.5023, \quad x(-2) = -1.9807.$$

Exercise 6.2.3

(a) Continue the preceding long division and obtain samples of $x(n)$ for the range $-3 \leq n \leq -10$.

(b) Use PC-DSP to check your hand calculations. The left-sided solution can be obtained by selecting the option *left-sided sequence* on the data-entry form for the inverse z-transform function.

Exercise 6.2.4

(a) A *stable* system has the z-domain system function

$$H(z) = \frac{z-1}{(z-0.7)(z-1.2)}.$$

Using the long-division method, determine the impulse response $h(n)$ of this system for the range of the sample index $-5 \leq n \leq 5$. Do this without computer aid. Note that the impulse response is a two-sided sequence in this case. First, separate $H(z)$ into two terms such that one term corresponds to a right-sided sequence and the other to a left-sided sequence. (Partial fraction expansion could be used for this.) Carry out long-division operation for each term and combine the results.

(b) Check your hand calculations using PC-DSP. Use the inverse z-transform function twice, once for the right-sided component of $X(z)$ and once for the left-sided component. Use the addition function accessed with *Operations/Arithmetic-operations/Add-sequences* to combine the results of the two steps.

6.3 DISCRETE-TIME STRUCTURES

In this section, we will briefly discuss the structures used for real-time implementation of discrete-time systems described by means of z-domain transfer functions. Consider a second-order DTLTI system with the system function

$$H(z) = \frac{z(z-1)}{z^2 - 1.4z + 1.13}.$$

The difference equation between the input signal $x(n)$ and the output signal $y(n)$ is

$$y(n) = 1.4y(n-1) - 1.13y(n-2) + x(n) - x(n-1).$$

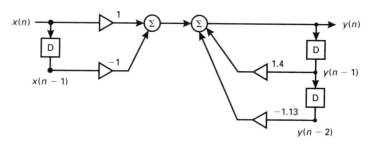

Figure 6.2 Direct-form type-1 block diagram for the example system.

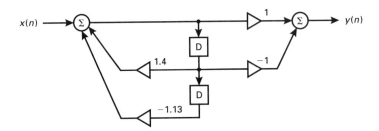

Figure 6.3 Direct-form type-2 block diagram for the example system.

A block diagram can be developed based on this difference equation as shown in Fig. 6.2. This is a called a *direct-form type-1* block diagram. Note that only three types of elements are used in this block diagram: adders, constant multipliers, and delays.

An alternative block diagram can be obtained by implementing the feedback portion of the system before the feed-forward portion. This leads to the *direct-form type-2* block diagram shown in Fig. 6.3.

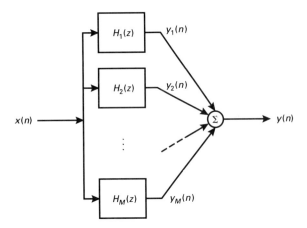

Figure 6.4 Parallel-form block diagram.

Higher-order systems are best implemented using parallel or cascade combinations of first-and second-order systems. If a system function can be written as a sum of lower-order system functions, that is,

$$H(z) = H_1(z) + H_2(z) + \cdots + H_M(z),$$

a parallel-form block diagram can be obtained as shown in Fig. 6.4. Alternatively, the system function can be written as a product of lower-order system functions, that is,

$$H(z) = H_1(z)H_2(z)\cdots H_M(z),$$

and a cascade-form block diagram can be obtained as shown in Fig. 6.5.

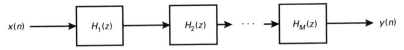

Figure 6.5 Cascade-form block diagram.

7

IIR Filter Design

The supreme guide in life is knowledge.
—Kemal Atatürk

In previous chapters, we considered analysis and implementation methods for infinite-impulse-response (IIR) filters. Linear and time-invariant IIR filters are typically described by means of a transfer function $H(z)$ or $H(\omega)$, an impulse response $h(n)$, or a difference equation specifying the relationship between the input and the output signals. Given an appropriate description of the filter in one of these three forms, its frequency-domain behavior can be observed in terms of magnitude, phase, time-delay, and group-delay characteristics or in terms of the placement of z-domain poles and zeros of the transfer function. A block diagram consisting of adders, multipliers, and delay elements can be developed by inspection of the transfer function, and the corresponding pseudocode can be written for a real-time implementation of the filter.

In this chapter, we will concentrate our efforts on the synthesis problem of developing an appropriate description (transfer function, impulse response, or difference equation) for an IIR filter that will behave in a prescribed way. In most cases, the desired filter behavior is specified in terms of one or several of the analysis characteristics mentioned and involves certain tolerance limits. If the design requirements can be satisfied at all, they can probably be satisfied by a number of different designs, so the answer to a design problem is not necessarily unique.

If a filter is being designed for real-time implementation, it must be stable and causal. Thus, stability and causality will be inherent requirements in all our design problems in this chapter.

7.1 OVERVIEW OF IIR FILTER DESIGN METHODS

The methods available for designing discrete-time IIR filters can be categorized in two major groups: (1) direct design methods and (2) indirect design methods based on analog filters.

Direct design of IIR filters is based on the idea of selecting a suitable form of the filter transfer function with unknown parameter values and computing the necessary parameters so as to minimize some cost function. As one particular example, let the desired filter magnitude behavior be given by $H_d(\omega)$. We might want to find a filter with the transfer function

$$H(\omega) = \frac{b_0 + b_1 e^{-j\omega}}{1 + a_1 e^{-j\omega} + a_2 e^{-j2\omega}}$$

as an approximation to the desired magnitude behavior. This can be accomplished by determining the coefficients b_0, b_1, a_1, and a_2 to minimize a *cost function* such as

$$J = \int_{-\pi}^{\pi} |H_d(\omega) - H(\omega)|^2 d\omega$$

which represents the mean-squared error of the approximation, subject to a scale factor. An alternative cost function to minimize might be

$$J = \int_{-\pi}^{\pi} |H_d(\omega) - H(\omega)|^p d\omega,$$

where p is any positive integer. It is also possible to incorporate a nonuniform weighting function into the cost function in order to emphasize some frequency ranges more than others. An example is

$$J = \int_{-\pi}^{\pi} W(\omega) |H_d(\omega) - H(\omega)|^2 d\omega$$

If it is desired to ensure that the approximation error is especially small in a certain range of frequencies, say $\omega_1 \leq \omega \leq \omega_2$, and if larger errors at other frequencies can be tolerated in return, then the weight function $W(\omega)$ can be chosen to have large amplitudes in the frequency range of interest and smaller values elsewhere.

Once a cost function is chosen, the coefficients of the filter are determined by minimizing it. To minimize J with respect to the parameters of $H(\omega)$, we need to differentiate it with respect to each parameter and set the result equal to zero; that is,

$$\frac{\partial J}{\partial b_0} = 0, \quad \frac{\partial J}{\partial b_1} = 0, \quad \frac{\partial J}{\partial a_1} = 0, \quad \frac{\partial J}{\partial a_2} = 0$$

In general, the resulting equations are nonlinear and difficult to solve. Solutions can only be found using iterative computer techniques that may or may not converge. Even if a solution is found, the filter obtained may or may not be stable. Because of this, we will not cover direct design of IIR filters any further in this text. Details of various direct design algorithms can be found in references 8, 18, and 24.

Indirect design methods for discrete-time IIR filters rely on well-established analog filter design methods. A suitable analog prototype filter is designed and then converted to a discrete-time filter. In the remainder of this chapter, we will concentrate on IIR design methods based on analog filter prototypes.

7.2 IIR FILTER SPECIFICATIONS

Before discussing IIR filter design methods, we need to establish the terminology and notational conventions used in describing the desired frequency domain behavior of IIR filters. Usually, the end result of the design procedure is the z-domain transfer function $H(z)$ of the designed filter, which can be evaluated on the unit circle of the z-plane to yield

$$H(\omega) = H(z)|_{z = e^{j\omega}}$$
$$= |H(\omega)|e^{j\arg[H(\omega)]}. \quad (7.1)$$

In problems involving the design of IIR filters from analog prototypes, filter specifications are usually given in terms of magnitude only. This does not mean that the phase response is unimportant, but it is simply due to the fact that well-developed design techniques do not exist for designing IIR filters from phase specifications.

We will first consider the specifications for a low-pass filter, depicted graphically in the specification diagram in Fig. 7.1. The magnitude response of the filter is required to stay within the unshaded tolerance limits. The range of frequencies from 0 to ω_1 radians is called the *passband* of the filter. In contrast, the range of frequencies from ω_2 to π radians is called the *stopband*. The parameters δ_1 and δ_2 are the *passband tolerance* and the *stopband tolerance*, respectively. Between the passband and the stopband, there is a third frequency band called a *transition band* (also called a *don't-care band*) in the range from ω_1 to ω_2 radians.

Note that, due to the periodic nature of the DTFT, the magnitude spectrum of any discrete-time filter is unique only for a 2π radian range of the angular frequency ω. If, in addition, the impulse response of the filter is real valued, then the magnitude spectrum is even. Thus, within a 2π radian range of ω, one half of the magnitude specification is redundant if the other half is specified. As shown in Fig. 7.1, the desired magnitude behavior is usually specified for the range $0 \leq \omega \leq \pi$.

The desired magnitude specifications can also be given in terms of the *normalized frequency* variable F, which is related to ω by the equation

$$\omega = 2\pi F. \quad (7.2)$$

The normalized frequency concept was introduced in Section 2.5 in the context of discrete-time sinusoids. If the independent variable ω extends from 0 to π radians, then the

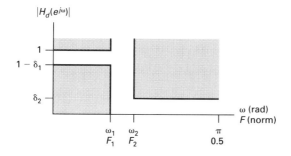

Figure 7.1 Magnitude specifications for a low-pass filter.

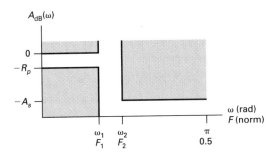

Figure 7.2 Log–magnitude specifications for a low-pass filter.

corresponding normalized frequency variable F extends from 0.0 to 0.5. Another way to look at normalized frequencies is to treat them as fractions of the sampling frequency. For example, if a 300-Hz continuous sinusoid is sampled with a sampling rate of 1 kHz, then the normalized frequency of the resulting discrete sinusoid is 0.3. Normalized frequencies are dimensionless quantities, since they are obtained through division of actual frequencies in Hertz by the sampling frequency, also in Hertz.

In designing discrete-time IIR filters using PC-DSP, filter specifications are entered using the normalized values of the edge frequencies involved. Thus, once the frequency axis is normalized, the solution of the design problem is independent of the actual value of the sampling frequency. This approach also makes it possible to have some uniformity in presenting the analysis results for the designed filter.

The log–magnitude response of the filter (in decibels) is obtained through the equation

$$A_{dB}(\omega) = 20 \log_{10}[|H(\omega)|]. \tag{7.3}$$

Magnitude specifications shown in Fig. 7.1 can be easily converted to log–magnitude specifications by taking the logarithm of vertical axis values and scaling by a factor of 20 (see Fig. 7.2). The maximum allowed decibel passband tolerance is defined as

$$\begin{aligned} R_p &= 20 \log_{10}(1) - 20 \log_{10}(1 - \delta_1) \\ &= 20 \log_{10}\left(\frac{1}{1 - \delta_1}\right), \end{aligned} \tag{7.4}$$

and the minimum required decibel stopband attenuation is

$$\begin{aligned} A_s &= 20 \log_{10}(1) - 20 \log_{10}(\delta_2) \\ &= 20 \log_{10}\left(\frac{1}{\delta_2}\right). \end{aligned} \tag{7.5}$$

The definitions presented for the low-pass filter of Figs. 7.1 and 7.2 apply to other filter types as well. Specification diagrams for high-pass, bandpass, and band-reject filters are given in Fig. 7.3.

Sec. 7.2　IIR Filter Specifications

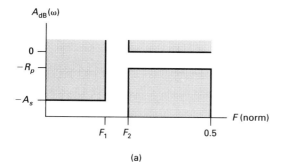

(a)

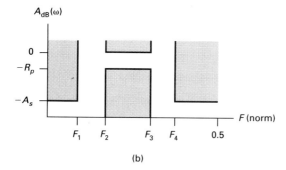

(b)

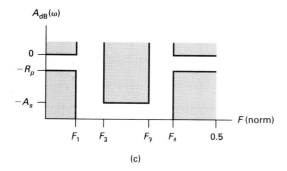

(c)

Figure 7.3 Specification diagrams for IIR filters: (a) High pass; (b) bandpass; (c) band-reject.

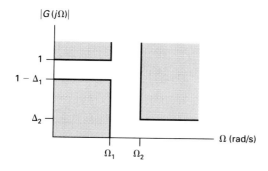

Figure 7.4 Specification diagram for an analog low-pass filter.

7.3 IIR FILTER DESIGN USING ANALOG PROTOTYPES

We are now ready to develop the details of the design procedure. We will assume that the desired filter is specified by means of a tolerance scheme as established in the previous section. In the case of low-or high-pass filters, the specifications consist of two critical frequencies ω_1 and ω_2 and the tolerance limits δ_1 and δ_2. If a bandpass or a band-reject filter is to be designed, additional critical frequencies ω_3 and ω_4 are also needed. Designing a discrete-time IIR filter by means of an analog prototype involves a three-step procedure:

1. The specifications of the desired discrete-time filter must be converted to the specifications of an appropriate analog filter that can be used as a prototype. Figure 7.4 shows the specification diagram for an analog low-pass filter that is very similar to that of a discrete-time low-pass filter. The critical frequencies Ω_1 and Ω_2 and the parameters Δ_1 and Δ_2 are determined based on ω_1, ω_2, δ_1, and δ_2.
2. An analog prototype filter that satisfies the specifications determined in step 1 is designed; that is, its transfer function $G(s)$ is found.
3. By means of an appropriate transformation, the analog prototype transfer function $G(s)$ is converted to a discrete-time filter transfer function $H(z)$. This completes the design. The filter designed can now be analyzed using the techniques presented in Chapters 3, 4, and 6. It can be implemented using the techniques of Chapter 6.

In a design problem, the three steps just outlined are performed in the order specified. In our treatment of IIR filter design, however, we will follow a different order. The problem of converting an analog transfer function into a discrete-time transfer function (step 3) will be covered first. Afterward, we will discuss the problem of obtaining analog prototype specifications from the discrete-time filter specifications (step 1). The reason for this change is due to the fact that the choice of which method to use in step 1 is based on which method is planned to be used in step 3. In order not to break the continuity, the problem of designing an analog prototype (step 2) will not be covered in this chapter and will be left to Appendix B instead.

7.4 ANALOG TO DISCRETE-TIME CONVERSION

Step 3 of the design procedure outlined in the previous section requires a discrete-time filter transfer function $H(z)$ to be obtained based on an analog prototype filter transfer function $G(s)$. In this step, the goal is to come up with a discrete-time transfer function $H(z)$ whose characteristics resemble, in some sense, those of the analog filter with transfer function $G(s)$. This is a rather loosely defined problem since we do not clearly define when two filters resemble each other and when they don't. Also, we do not state which characteristics of the two filters will resemble each other (that is, impulse response, magnitude spectrum, delay characteristics, and so on). In most cases, analog prototype filters are designed to satisfy certain requirements in terms of their magnitude spectra, and we are therefore usually interested in obtaining a reasonable match between the magnitude characteristics of the two filters.

Sec. 7.4 Analog to Discrete-time Conversion

Recall from Chapter 3 that the system function $H(\omega)$ of any discrete-time filter must be periodic with period 2π radians. On the other hand, the system function $G(\Omega)$ of a continuous-time analog prototype filter does not have this property. Thus, an exact match between the two transfer functions is not possible, and approximate methods must be used. In this section, we will discuss four such methods: impulse invariance, first backward difference, bilinear transformation, and matched z-transform.

Impulse Invariance

One possible method of obtaining a discrete-time filter from an analog prototype filter is to preserve the impulse response in the conversion process. Specifically, the discrete-time filter is chosen such that its impulse response is a sampled version of the analog prototype filter impulse response, that is,

$$h(n) = Tg(nT) \tag{7.6}$$

where $h(n)$ and $g(t)$ are the impulse responses of the discrete-time filter and the analog prototype, respectively. The relationship given by (7.6) is the sampling relationship discussed in Chapter 5. As a result of this, the transfer function of the discrete-time filter is related to the analog prototype transfer function by

$$H(\omega) = \sum_{k=-\infty}^{\infty} G\left(\frac{\omega + 2\pi k}{T}\right). \tag{7.7}$$

Assume that the analog prototype filter is band limited such that

$$G(\Omega) = 0, \quad |\Omega| > \Omega_{max} \tag{7.8}$$

and $\Omega_{max} \leq \frac{\pi}{T}$ to satisfy the Nyquist sampling criterion. In this case, we can write

$$H(\omega) = G\left(\frac{\omega}{T}\right), \quad -\pi \leq \omega \leq \pi. \tag{7.9}$$

Clearly, the discrete-time filter transfer function is a scaled version of the analog filter transfer function with $\Omega = \omega/T$. Nevertheless, the band-limiting condition given by (7.8) is usually not satisfied by the analog prototype filter. In this case, the resulting discrete-time filter spectrum is an aliased version of the analog prototype spectrum. (See Fig. 7.5.) This limits the application of the impulse invariance technique to low-pass and some bandpass filters where effects of aliasing can be tolerated.

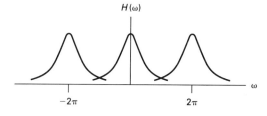

Figure 7.5 Aliasing effect in impulse invariant filter design.

A rational transfer function $G(s)$ can be expanded into partial fractions to yield

$$G(s) = \sum_{i=1}^{M} \frac{A_i}{s - s_i} \quad (7.10)$$

where s_i represent poles of the analog prototype filter and A_i are the corresponding residues. The impulse response of the filter can be found by computing the inverse Laplace transform of (7.10):

$$g(t) = \sum_{i=1}^{M} A_i e^{s_i t} u(t). \quad (7.11)$$

We will sample the impulse response $g(t)$ to obtain the impulse response of the discrete-time filter.

$$h(n) = Tg(t)\big|_{t=nT}$$

$$= \sum_{i=1}^{M} T A_i e^{s_i nT} u(n). \quad (7.12)$$

Finally, taking the z-transform of (7.12) yields

$$H(z) = \sum_{i=1}^{M} \frac{T A_i z}{z - e^{s_i T}} \quad (7.13)$$

A comparison of (7.10) and (7.13) reveals that it is possible to obtain $H(z)$ from the partial fraction expansion of $G(s)$ without the intermediate steps of obtaining $g(t)$ and sampling it.

An important issue in analog to discrete-time filter conversion is the stability of the resulting discrete-time filter. The poles of the analog prototype filter are at

$$s_i = \sigma_i + j\Omega_i, \quad i = 1, \ldots, M.$$

The corresponding poles of the discrete-time filter are at

$$z_i = e^{s_i T}$$
$$= e^{\sigma_i T} e^{j\Omega_i T}, \quad i = 1, \ldots, M.$$

If the analog prototype filter is stable, its poles are in the left half of the s-plane, and thus $\sigma_i < 0$ for all poles. This implies $|z_i| < 1$ for the poles of the discrete-time filter. Thus, if the analog prototype filter is stable, the discrete-time filter obtained through impulse invariance will also be stable.

Exercise 7.4.1

An analog low-pass filter is described by the rational transfer function

$$G(s) = \frac{60.9344}{(s^2 + 3.9351s + 15.4849)(s + 3.9351)}$$

(a) This analog filter has already been saved into a PC-DSP compatible data file under the name EX-7-4-1.FLT and is on your distribution disks. Using PC-DSP, compute and graph the magnitude and the phase characteristics of the filter. Note that this can be accomplished by

Sec. 7.4 Analog to Discrete-time Conversion

using the menu selections *Filters/Analyze-filter*, specifying the name of the filter data file, and selecting the analysis type desired (that is, magnitude, phase, or the like).

(b) Using the impulse invariance technique outlined with a sampling interval $T = 1$ s, find the z-domain transfer function of a corresponding discrete-time filter. Do this part manually.

(c) PC-DSP can be used to perform part (b). Use the menu selections *Filters/Transform-analog-filter*, and select *Impulse-invariance* in the data-entry window presented. Specify the name of the analog filter used as input and the name of the discrete-time filter desired. Compare the resulting discrete-time filter to your hand calculations in part (a). Compute and graph the magnitude and the phase characteristics of the discrete-time filter obtained. Is the aliasing effect noticeable?

(d) Repeat part (c) with a sampling interval of $T = 0.5$ s. Compare the magnitude response of the filter to that obtained in part (c). How was the cutoff frequency of the resulting discrete-time filter affected by the change in T? How significant is aliasing compared to part (c)?

The preceding exercise should demonstrate the fact that reducing the sampling interval by half (thus doubling the sampling rate) reduces aliasing, but it also results in a proportional change in the cutoff frequency of the discrete-time filter. Recall that analog frequencies and discrete-time (angular) frequencies are related by $\Omega = \omega/T$. In a typical design problem, the cutoff frequency of the discrete-time filter is specified, and the cutoff frequency of the analog prototype filter is computed accordingly. Therefore, the ratio of the analog filter cutoff frequency to the sampling rate is fixed. Let the cutoff frequencies for the analog filter and the impulse invariant discrete-time filter be Ω_c and ω_c, respectively. The impulse response of the analog filter is sampled at the rate

$$\Omega_s = \frac{2\pi}{T}.$$

Hence

$$\frac{\Omega_c}{\Omega_s} = \frac{\omega_c}{2\pi} = \text{constant}.$$

For a given discrete time cutoff frequency, increasing the sampling rate does not improve the aliasing condition, since the bandwidth of the signal being sampled also increases proportionally. In this sense, the choice of T is irrelevant, and $T = 1$ s is typically used for simplicity.

Exercise 7.4.2

An analog high-pass filter is described by the rational transfer function

$$G(s) = \frac{s^3}{(s^2 + 2.5081s + 6.2906)(s + 2.5081)}.$$

(a) This analog filter has already been saved into a PC-DSP compatible data file under the name EX-7-4-2.FLT and is on your distribution disks. Using PC-DSP, compute and graph the magnitude and the phase characteristics of the filter.

(b) Again using PC-DSP, transform this filter into a discrete-time filter through impulse invariant transformation. Use the sampling interval $T = 1$ s. Compute and graph the magnitude and the phase characteristics of the resulting discrete-time filter. Is the result satisfactory? Explain.

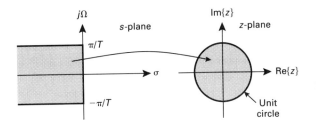

Figure 7.6 Mapping between the *s*-plane and *z*-plane for the impulse-invariant transformation.

For a DTLTI system, magnitude and phase characteristics are obtained by evaluating the system function $H(z)$ on the unit circle of the *z*-plane. Similarly, for a continuous-time linear and time-invariant system, these characteristics are obtained by evaluating the system function $G(s)$ on the $j\Omega$-axis of the *s*-plane. Figure 7.6 illustrates the mapping between the *s*-plane and the *z*-plane for the impulse-invariant transform. Note that consecutive sections of length $2\pi/T$ on the $j\Omega$-axis of the *s*-plane map onto the unit circle of the *z*-plane. Thus, the magnitude (or phase) response of $H(z)$ on the unit circle of the *z*-plane is the sum of magnitude (or phase) responses of an infinite number of sections of $G(s)$ of the $j\Omega$-axis of the *s*-plane. Since the mapping is not one-to-one, aliasing results.

First Backward Difference

An analog prototype filter can also be described by a linear constant-coefficient differential equation. Another method of obtaining a discrete-time filter from an analog prototype is to convert this differential equation into a corresponding difference equation by approximating each differential with a first backward difference.

Consider a first-order analog filter with the transfer function

$$G(s) = \frac{Y_a(s)}{X_a(s)} = \frac{a}{s - b}. \tag{7.14}$$

The corresponding differential equation is

$$\frac{dy_a(t)}{dt} = by_a(t) + ax_a(t).$$

At time instant $t = nT$, we can write

$$\left.\frac{dy_a(t)}{dt}\right|_{t = nT} = by_a(nT) + ax_a(nT). \tag{7.15}$$

Let two sequences be defined as $y(n) = y_a(nT)$ and $x(n) = x_a(nT)$. If we use the first backward difference (see Exercise 3.4.8) to approximate the derivative on the left side of (7.15), the corresponding difference equation is

$$y(n)\left[\frac{1}{T} - b\right] = \frac{1}{T}y(n - 1) + ax(n),$$

Sec. 7.4 Analog to Discrete-time Conversion

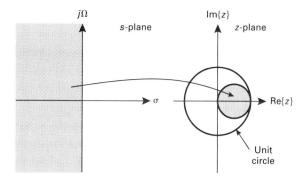

Figure 7.7 Mapping between the s-plane and the z-plane for the first backward difference technique.

which leads to a discrete-time filter with the z-domain transfer function

$$H(z) = \frac{a}{[(1 - z^{-1})/T] - b}. \qquad (7.16)$$

Comparing (7.14) and (7.16), the transformation from the s-plane to the z-plane is obvious:

$$s = \frac{1 - z^{-1}}{T}. \qquad (7.17)$$

Even though we have derived the transformation in (7.17) for a first-order analog prototype filter, it applies to rational transfer functions of any order. A higher-order transfer function can be expanded into partial fractions for which each term in the expansion is of the type given by (7.14). The transformation in (7.17) can then be used on each term. *Note:* This explanation is not mathematically rigorous since it does not cover the possibility of multiple-order poles. The transformation in (7.7) is valid even for systems with multiple-order poles; however, the proof will not be given here.

It can also be shown that stable analog filters lead to stable discrete-time filters; that is, if all poles of the analog prototype are in the left half of the s-plane, then the corresponding poles of the discrete-time filter are inside the unit circle of the z plane.

In (7.17), if the parameter s is varied on the $j\Omega$-axis of the s-plane, corresponding values of z lie on a circle with radius ½ in the z-plane. The center of the circle is at $z = $ ½. Points in the left half s-plane are mapped to points inside this little circle in the z-plane. This relationship is illustrated in Fig. 7.7. The trajectory circle is tangent to the unit circle of the z-plane at the point $z = 1$, which corresponds to the angular frequency $\omega = 0$. As a result of this, the similarity between analog and discrete-time filter frequency responses is usually good at low frequencies, but is degraded at higher frequencies, effectively limiting the use of the first backward difference technique to low-pass and some bandpass filter designs. The degradation of the frequency response at high frequencies can also be explained from a time-domain perspective. In the differential equation of the analog prototype filter, derivatives are approximated with backward differences using a fixed step size T. Naturally, the quality of this is approximation will be better for slowly varying signal components (low frequencies) compared to rapidly varying signal components (high frequencies).

Exercise 7.4.3

(a) Consider again the analog filter transfer function used in Exercise 7.4.1. Using the first backward difference technique with sampling interval $T = 1$ s, find the z-domain transfer function of a corresponding discrete-time filter. Do this part manually.

(b) PC-DSP can be used to do part (a). Use the menu selections *Filters/Transform-analog-filter* and select *First-backward-difference* in the data-entry window presented. Specify the name of the analog filter used as input and the name of the discrete-time filter desired. Compute and graph the magnitude and the phase characteristics of the discrete-time filter obtained. Observe the degree of similarity between the analog and discrete-time magnitude responses at low and high frequencies.

(c) Repeat part (b) with a sampling interval of $T = 0.5$ s. Compare the magnitude response of the filter to that obtained in part (b). How does this change affect the degree of similarity between the analog and discrete-time magnitude responses?

Exercise 7.4.4

Using the first backward difference technique on the analog high-pass filter transfer function given in Exercise 7.4.2, attempt to obtain a discrete-time high-pass filter. Use the sampling interval $T = 1$ s. Compute and graph the magnitude and the phase characteristics of the resulting discrete-time filter. Is the result satisfactory? Explain.

Exercise 7.4.5

An analog low-pass elliptic filter has the transfer function

$$G(s) = \frac{60.9344}{(s^2 + 3.9351s + 15.4849)(s + 3.9351)}.$$

We will use this filter as a prototype to observe the degradation of the discrete-time filter frequency response at high frequencies when the first backward difference method is used.

(a) Compute and plot the magnitude and phase characteristics of the analog filter.

(b) Using the first backward difference technique with $T = 1$ s, find the corresponding discrete-time filter. Compute and plot its magnitude and phase characteristics. Roughly determine the range of frequencies in which analog and discrete-time filter responses look "similar."

(c) Repeat part (b) with $T = 0.5$ s. Comment on the results.

Bilinear Transformation

The two transformation methods discussed so far have severe shortcomings that prevent them from being adopted as general methods applicable for all types of filters. In the case of the impulse-invariant transformation, the aliasing effect that results from sampling the impulse response limits its use to the design of low-pass and some bandpass filters. When the first backward difference technique is used, aliasing is avoided, but the frequency response of the resulting discrete-time filter does not closely match that of the analog prototype, especially for high frequencies. We will see shortly that bilinear transformation alleviates these problems.

Consider a first-order analog prototype filter described by the transfer function

Sec. 7.4 Analog to Discrete-time Conversion

$$G(s) = \frac{Y_a(s)}{X_a(s)} = \frac{a}{s-b}. \tag{7.18}$$

(Later, we will see that the techniques developed in this section will be applicable to higher-order filters as well.) The differential equation that describes the time-domain behavior of this filter is

$$\frac{dy_a(t)}{dt} = by_a(t) + ax_a(t). \tag{7.19}$$

Let's define an intermediate variable $w_a(t)$ as

$$w_a(t) = \frac{dy_a(t)}{dt}. \tag{7.20}$$

Equation (7.19) can now be written as

$$w_a(t) = by_a(t) + ax_a(t). \tag{7.21}$$

With the discrete sequence definitions $w(n) = w_a(nT)$, $x(n) = x_a(nT)$, and $y(n) = y_a(nT)$, we can write a discrete-time version of (7.21); that is,

$$w(n) = by(n) + ax(n). \tag{7.22}$$

On the other hand, integrating both sides of (7.20), we obtain

$$y_a(t) = \int_{t_0}^{t} w_a(t)\, dt + y_a(t_0). \tag{7.23}$$

Letting $t_0 = nT - T$ and using the trapezoidal approximation method (see Exercise 3.4.5) for numerically approximating the integral in (7.23),

$$y(n) = \frac{T}{2}[w(n) + w(n-1)] + y(n-1). \tag{7.24}$$

Finally, substituting (7.22) into (7.24), the difference equation between $x(n)$ and $y(n)$ is

$$y(n) = \frac{T}{2}[by(n) + ax(n) + by(n-1) + ax(n-1)] + y(n-1). \tag{7.25}$$

The transfer function of the corresponding discrete-time filter is obtained by taking the z-transform of both sides of (7.25) and rearranging terms:

$$H(z) = \frac{Y(z)}{X(z)} \tag{7.26}$$

$$= \frac{a}{\dfrac{2}{T}\dfrac{1-z^{-1}}{1+z^{-1}} - b}.$$

By comparing (7.18) and (7.26), we see that the transformation between the s-plane and the z-plane is given by

$$s = \frac{2}{T}\cdot\frac{1-z^{-1}}{1+z^{-1}}. \tag{7.27}$$

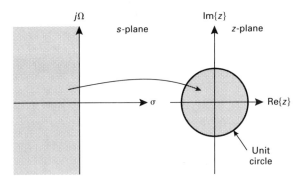

Figure 7.8 Mapping between the s-plane and z-plane for the bilinear transformation technique.

This is known as the bilinear transformation. Solving (7.27) for z, we get

$$z = \frac{2 + sT}{2 - sT} \tag{7.28}$$

as the inverse transformation. Thus, bilinear transformation represents a one-to-one mapping of the s-plane onto the z-plane; that is, for each point in the s-plane, there is one and only one point in the z-plane, and vice versa. To find out how the $j\Omega$-axis of the s-plane maps onto the z-plane, (7.28) can be evaluated for $s = j\Omega$, yielding

$$z = \frac{2 + j\Omega T}{2 - j\Omega T}.$$

It can be easily shown that the magnitude of z as given by this expression is equal to unity, independent of the value of Ω. Thus, the $j\Omega$-axis of the s-plane maps onto the unit circle of the z-plane. Since the entire $j\Omega$-axis of the s-plane is mapped onto the unit circle of the z-plane only once, there is no aliasing. Therefore, unlike the transformation techniques previously covered, bilinear transformation is not limited to low-pass filters and can be used for the design of all four frequency-selective filter types.

The relationship between analog and discrete-time (angular) frequencies is

$$\omega = 2 \tan^{-1}\left(\frac{\Omega T}{2}\right). \tag{7.29}$$

This relationship is shown graphically in Fig. 7.8. The frequency axis of the analog filter is warped (distorted) in the process of mapping the s-plane onto the z-plane. This is unavoidable since an infinite range of frequencies in the s-plane is mapped into a 2π-radian range of angular frequencies in the z-plane, and the mapping is one to one. It is possible, however, to compensate for this distortion in converting the specifications of the desired discrete-time filter to the specifications of the analog prototype. This will be covered in the next section.

Exercise 7.4.6

(a) Convert the analog filter transfer function used in Exercise 7.4.1 into a discrete-time filter transfer function using the bilinear transformation technique. Specifically, substitute (7.27) into the s-domain transfer function, and simplify the resulting z-domain transfer function. Do this part manually. Use $T = 1$ s.

Sec. 7.4 Analog to Discrete-time Conversion

(b) PC-DSP can be used to do part (a). Use the menu selections *Filters/Transform-analog-filter*, and select *Bilinear-transformation* in the data-entry window presented. Specify the name of the analog filter used as input and the name of the discrete-time filter desired. Compute and graph the magnitude and phase characteristics of the discrete-time filter obtained. Compare the results to those obtained earlier through the use of impulse invariance and first backward difference techniques.

(c) Change T to 0.5 s and repeat part (b). How does this change affect the mapping of analog frequencies to discrete-time frequencies? Does it correspond to a proportional scaling of the frequency axis?

Exercise 7.4.7

Convert the analog high-pass filter transfer function given in Exercise 7.4.2 into a discrete-time high-pass filter using the bilinear transformation technique. Use $T = 1$ s. Compute and graph the magnitude and phase characteristics of the resulting discrete-time filter. Is the result satisfactory? Can bilinear transformation be used for designing high-pass filters as well? Explain.

Exercise 7.4.8

(a) Consider again the analog low-pass elliptic filter used in Exercise 7.4.5. Using PC-DSP, compute and graph the magnitude response of this filter. Record the frequencies at which the maxima and minima of the magnitude response occur. (You might want to tabulate the magnitude response to determine these frequencies more accurately.)

(b) Convert this analog filter into a discrete-time filter using bilinear transformation with $T = 1$ s. Compute and graph the magnitude response of this discrete-time filter. Determine the frequencies of the maxima and minima. Verify Eq. (7.29) for corresponding analog and discrete-time frequency pairs.

Matched *z*-Transform

The last method we will discuss for converting an analog prototype filter into a discrete-time filter is the matched *z*-transform. This is a simple technique that is based on mapping *s*-domain poles and zeros of the analog prototype filter directly into the *z*-domain and constructing a *z*-domain transfer function using these poles and zeros. Consider an analog prototype filter transfer function in the form

$$G(s) = A \frac{\prod_i (s - \alpha_i)}{\prod_j (s - \beta_j)}.$$

A discrete-time filter can be constructed in the form

$$H(z) = \tilde{A} \frac{\prod_i (z - e^{\alpha_i T})}{\prod_j (z - e^{\beta_j T})}.$$

The mapping from the *s*-domain to the *z*-domain ensures a stable discrete-time filter as long as the analog prototype filter is stable. Note that the poles obtained using this technique are identical to those obtained using the impulse-invariance technique. The locations of zeros, however, are different. Aliasing considerations and filter-type limitations mentioned in the discussion of the impulse-invariance technique apply here as well.

Exercise 7.4.9

(a) Manually convert the analog low-pass filter used in Exercise 7.4.1 into a discrete-time filter using the matched z-transform technique. Use $T = 1$ s. In addition to mapping the poles of the transfer function, adjust the constant gain factor such that the resulting discrete-time filter has unit magnitude at $\omega = 0$.

(b) Repeat part (a) using PC-DSP. Use the menu selections *Filters/Transform-analog-filter*, and select *Matched-z-transform* in the data-entry window presented. Specify the name of the analog filter used as input and the name of the discrete-time filter desired. Compare the resulting discrete-time filter to your hand calculations in part (a). Compute and graph the magnitude and phase characteristics of the discrete-time filter obtained. Is the aliasing effect noticeable?

(c) Repeat part (b) with a sampling interval of $T = 0.5$ s. Compare the magnitude response of the filter to that obtained in part (b). How was the cutoff frequency of the resulting discrete-time filter affected by the change in T? How significant is aliasing compared to part (b)?

7.5 OBTAINING ANALOG PROTOTYPE SPECIFICATIONS

As discussed in Section 7.3, the first step in designing analog-prototype-based IIR filters is to convert the specifications of the desired discrete-time filter to the corresponding specifications of an appropriate analog prototype filter. For low-and high-pass filters, the desired magnitude behavior may be specified in terms of the critical frequencies ω_1 and ω_2, along with passband and stopband tolerance values δ_1 and δ_2. Corresponding parameters of the analog prototype filter are Ω_1, Ω_2, Δ_1, and Δ_2. For bandpass and band-reject filters, two more critical frequencies need to be specified.

Passband and stopband tolerance values of the discrete-time filter are directly translated to the corresponding tolerance values of the analog prototype; that is,

$$\Delta_i = \delta_i \tag{7.30}$$

for $i = 1, \ldots, 4$. On the other hand, frequency specifications must be translated based on which method is to be used in step 3 of the design process. If impulse invariance or the matched z-transform will be used in converting the analog prototype into a discrete-time filter, critical frequencies of the analog prototype are computed as

$$\Omega_i = \frac{\omega_i}{T} \tag{7.31}$$

since the conversion process essentially amounts to sampling the impulse response of the analog prototype to obtain the impulse response of the discrete-time filter. If bilinear transformation is to be used in converting the analog prototype into a discrete-time filter, the distortion of the frequency axis must be taken into account. In this case, critical frequencies of the analog prototype are computed as

$$\Omega_i = \frac{2}{T} \tan\left(\frac{\omega_i}{2}\right) \tag{7.32}$$

This is called *prewarping*. Critical frequencies are warped, that is, distorted, using (7.32). Recall that, when bilinear transformation is applied to the analog prototype filter in step

Sec. 7.6 Designing IIR Filters Using PC-DSP

3 of the design process, the entire frequency axis is distorted as described by (7.29). Thus, prewarping counteracts this distortion at the critical frequencies.

If the first backward difference technique is to be used in step 3, then the translation of critical frequencies is a somewhat awkward problem. Recall that this technique maps the $j\Omega$-axis of the s-plane onto a trajectory other than the unit circle of the z-plane. Specifically, this trajectory is a circle centered at $(\frac{1}{2} + j0)$ with a radius of $\frac{1}{2}$. On the other hand, angular frequencies are defined only on the unit circle of the z-plane. Solving (7.17) for s and evaluating the solution on the $j\Omega$-axis of the s-plane, we obtain

$$z = \frac{1}{1 - j\Omega T}. \tag{7.33}$$

If $\Omega T \ll 1$, we can write

$$\frac{1}{1 - j\Omega T} \approx 1 + j\Omega T \approx e^{j\Omega T}. \tag{7.34}$$

Thus, critical frequencies can be translated using

$$\Omega_i = \frac{\omega_i}{T}$$

provided that $\Omega_i T \ll 1$ or, equivalently, $\omega_i \ll 1$. This is consistent with the fact that the magnitude behavior of the discrete-time filter resembles that of the analog prototype filter only at low frequencies for which the two circles are close to each other.

7.6 DESIGNING IIR FILTERS USING PC-DSP

In this section, a number of IIR filter design exercises will be presented. PC-DSP has an IIR filter design section accessed through menu selections *Filters/IIR-filter-design*. It facilitates quick design of IIR filters using the bilinear transformation technique. In addition, it is also possible to design an IIR filter by using separate steps, such as designing an analog low-pass filter, applying a frequency transformation as necessary, and converting it to a discrete-time filter using one of the four conversion techniques discussed previously. This latter approach involves several steps and is somewhat more tedious than the former, but it is also more instructive and should be preferred if the goal is to learn IIR filter design.

Exercise 7.6.1

An IIR low-pass filter is to be designed to satisfy the following specifications:

$$0.99 \leq |H(\omega)| \leq 1, \quad \text{for } 0 \leq \omega \leq 0.4\pi,$$
$$0 \leq |H(\omega)| \leq 0.01, \quad \text{for } 0.6\pi \leq \omega \leq \pi.$$

The design procedure will be based on finding an appropriate Butterworth analog filter and converting it to a discrete-time filter using the impulse-invariance technique. The lowest-order filter that satisfies the specifications is desired.

(a) Find the specifications (critical frequencies and tolerance limits) for the analog prototype filter. Do this part manually.

(b) Determine the lowest filter order necessary to satisfy the specifications found in part (a). Use the analog filter design formulas derived in Appendix B. No computer use should be necessary for this part.

(c) Using PC-DSP, design the analog prototype filter. Use menu selections *Filters/Analog-filter-design*. Answer the prompts for filter type, approximation type, critical frequencies, and the name of the filter.

(d) Analyze the analog prototype filter designed in part (c). Use menu selections *Filters/Analyze-filter*. Compute and graph the magnitude response. Check to see if it satisfies the tolerance limits specified.

(e) Convert the analog filter designed in part (c) to a discrete-time filter using the impulse-invariance technique. Use menu selections *Filters/Transform-analog-filter* and select *Impulse-invariance*. Use the sampling interval $T = 1$ s. Analyze the resulting discrete-time filter in terms of magnitude and phase responses and the locations of its poles and zeros.

Exercise 7.6.2

Repeat the requirements of Exercise 7.6.1 using the following:

(a) First backward difference method.

(b) Matched z-transform method.

(c) Bilinear transformation method.

Note that, for first backward difference and matched z-transform techniques, only part (e) of Exercise 7.6.1 needs to be done. For bilinear transformation, however, part (a) of Exercise 7.6.1 also needs to be repeated since critical frequencies should be prewarped.

Exercise 7.6.3

(a) Design a Chebyshev type 2 band-reject filter to remove the normalized frequency $F = 0.22$. Choose the stopband edge frequencies such that the stopband is centered around the normalized frequency 0.22 and its width is 0.04. Choose the widths of both transition bands to be equal to 0.03. The maximum variation allowed for the passband magnitude response is 1 dB. At least 20-dB attenuation must be ensured in the stopband.

(b) Compute and graph the impulse-response, magnitude, phase, time-delay, and group-delay characteristics of the designed filter.

(c) View the design report for the filter. This report is accessed by pressing the button labeled *Report* while in the *Analyze-filter* section of PC-DSP.

(d) Tabulate the transfer function coefficients of the designed filter. Note that the z-domain transfer function of the filter is expressed in terms of first-and second-order cascade sections; that is,

$$H(z) = A \prod H_i(z),$$

where each section has the transfer function

$$H_i(z) = \frac{a_{i,2}z^2 + a_{i,1}z + a_{i,0}}{z^2 + b_{i,1}z + b_{i,0}}.$$

The coefficients of each stage and the filter gain factor are tabulated. Using the coefficients obtained, draw a cascade-form block diagram and write the corresponding pseudocode for the filter.

(e) Generate 300 samples of the sequence

$$x(n) = \sin[2\pi(0.22)n], \quad n = 0, \ldots, 299,$$

and process with the band-reject filter designed. In PC-DSP, this can be achieved using menu selections *Filters/Simulate-filter* and specifying the names of the filter and the input signal. Tabulate and graph the resulting output sequence and comment on it. Does the filter start attenuating the sinusoid immediately or is there a transition period?

Exercise 7.6.4

A discrete-time linear time-invariant system is described by the z-domain transfer function

$$H(z) = \frac{z}{z - 0.9}.$$

(a) Using menu selections *Filters/Enter-external-filter*, create a PC-DSP compatible description of this system as an IIR filter. The transfer function can be entered either in terms of numerator and denominator coefficients or in terms of z-domain pole and zero locations. Once entered into PC-DSP, the filter can be analyzed just like any other filter designed using PC-DSP.

(b) Compute and plot the magnitude and phase characteristics of the frequency response. Note the magnitude and phase at the normalized frequency 0.05.

(c) Generate 500 samples of the sinusoid

$$x(n) = \sin[2\pi(0.05)n], \quad n = 0, \ldots, 499$$

and process through the filter described in part (a). Compare the steady-state portion of the response to the original input sequence. How are the amplitude and phase of the sinusoid affected by the filter?

(d) Repeat part (c) with the sinusoid

$$x(n) = \sin[2\pi(0.1)n], \quad n = 0, \ldots, 499.$$

8

FIR Filter Design

> Discovery consists in seeing what everyone else has seen and thinking what no one else has thought. —Albert Szent-Gyorgi

In this chapter, the problem of designing finite-impulse-response (FIR) filters will be discussed. An FIR filter is completely characterized by an impulse response $h(n)$ that is nonzero in the range $0 \leq n \leq N - 1$. In a typical FIR filter design problem, the desired ideal frequency response is given. The design procedure consists of the following steps:

1. Based on the desired frequency response, select an appropriate value for the number of filter coefficients N. This might be just an educated guess. In some cases, empirical formulas are available.
2. Choose a design method that attempts to minimize, in some sense, the difference between the desired frequency response and the frequency response obtained.
3. Using the design method chosen, determine the set of filter coefficients $h(n)$.
4. Analyze the designed filter and decide if it is satisfactory. If it is not, then repeat the design with a different value of N and/or a different design method.

Recall that, in Chapter 4, we established the conditions under which the phase characteristic of an FIR filter becomes a linear function of the angular frequency. Usually, the possibility of linear-phase behavior provides the motivation for designing FIR filters.

Sec. 8.1 Fourier Series Design of FIR Filters

As we saw in Chapter 6, when real-time implementation of digital filters is concerned, IIR filters are mathematically more efficient and less demanding of hardware resources. If linear phase is not a requirement in a filtering application, then an IIR filter is usually preferred. Therefore, linear phase will be an inherent requirement in all FIR filter design methods discussed in this chapter.

First, the Fourier series design technique will be developed. In this technique, the desired ideal filter behavior is equal to either unity or zero at a given frequency. The use of window functions to remedy the fundamental problems of this technique will also be discussed. Afterward, we will discuss the least-squares design method where a transition band is added into the specifications, and the desired behavior in the transition band is also specified. Finally, the Parks–McClellan technique will be discussed. In this technique, the specifications of the desired filter also include a transition band; however, the desired response is not specified in the transition band.

Sometimes, FIR filters are designed to approximate ideal differentiators or ideal Hilbert transform filters. An ideal differentiator has the system function

$$H(\omega) = j\omega, \tag{8.1}$$

which can also be written as

$$H(\omega) = \omega e^{j\pi/2}. \tag{8.2}$$

Note that (8.2) is given for the range $-\pi < \omega < \pi$ and is repeated periodically outside this range. The system function for an ideal Hilbert transform filter is

$$H(\omega) = j\,\text{sgn}(\omega)$$
$$= \begin{cases} j, & 0 < \omega < \pi \\ -j, & -\pi < \omega < 0 \end{cases}$$

which is also periodic with period 2π.

8.1 FOURIER SERIES DESIGN OF FIR FILTERS

An ideal low-pass filter with a cutoff frequency ω_c is characterized by a system function in the form

$$H_d(\omega) = \begin{cases} 1, & |\omega| < \omega_c \\ 0, & \omega_c < |\omega| < \pi, \end{cases} \tag{8.3}$$

which is depicted graphically in Fig. 8.1. Recall that any sequence $x(n)$ can be thought of as a linear combination of sinusoidal basis functions. If $x(n)$ is the input to this filter, the output signal $y(n)$ contains only those frequency components of $x(n)$ that are in the range $-\omega_c < \omega < \omega_c$. Using the inverse discrete Fourier transform, the impulse response of this ideal filter is

$$h_d(n) = \frac{1}{2\pi} \int_{-\pi}^{\pi} H_d(\omega) e^{j\omega n}\, d\omega$$

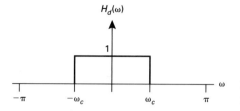

Figure 8.1 Ideal low-pass filter frequency response.

$$= \frac{1}{2\pi} \int_{-\omega_c}^{\omega_c} e^{j\omega n} \, d\omega \qquad (8.4)$$

$$= \frac{\sin(\omega_c n)}{\pi n}.$$

This impulse response is shown in Fig. 8.2.

Unfortunately, (8.4) is valid for all n in the range $-\infty < n < \infty$; that is, the impulse response $h_d(n)$ that we just found is of infinite length and thus cannot be the impulse response of an FIR filter. On the other hand, its samples seem to get smaller in amplitude as the index n grows in either direction. A finite-length sequence can be obtained by truncating the infinite-length sequence $h_d(n)$ to $2M + 1$ samples.

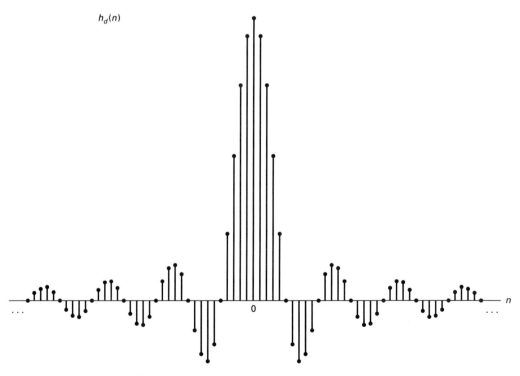

Figure 8.2 Ideal low-pass filter impulse response.

Sec. 8.1 Fourier Series Design of FIR Filters

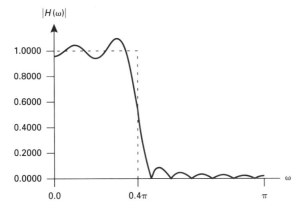

Figure 8.3 Magnitude characteristic of a 21-tap FIR low-pass filter with $\omega_c = 0.4\pi$ radians.

$$h_T(n) = \begin{cases} h_d(n), & -M \leq n \leq M \\ 0, & \text{otherwise.} \end{cases} \quad (8.5)$$

The system function for the resulting filter is

$$H_T(\omega) = \sum_{n=-\infty}^{\infty} h_T(n) e^{-j\omega n} \quad (8.6)$$

$$= \sum_{n=-M}^{M} h_d(n) e^{-j\omega n}.$$

Note that the impulse response $h_T(n)$ is noncausal since it has nonzero samples for $-M \leq n \leq -1$. To obtain a causal filter, a delay of M samples should be incorporated into the impulse response; that is,

$$h(n) = h_T(n - M). \quad (8.7)$$

The system function of the resulting FIR filter is

$$H(\omega) = e^{-j\omega M} H_T(\omega). \quad (8.8)$$

The M-sample delay affects only the phase of the resulting system function and not its magnitude. Figure 8.3 shows the magnitude response of a 21-tap ($M = 10$) FIR low-pass filter with cutoff frequency $\omega_c = 0.4\pi$ radians. Note the oscillatory behavior of the magnitude especially around the cutoff frequency.

Exercise 8.1.1

(a) First, without the aid of the computer, design a nine-tap FIR low-pass filter with $\omega_c = 0.4\pi$ radians. Using (8.4), (8.5), and (8.7) along with the parameter values given, the impulse response of the filter can be written in the compact form (verify)

$$h(n) = \frac{\sin[0.4\pi(n-4)]}{\pi(n-4)}, \quad n = 0, \ldots, 8.$$

Evaluate samples of $h(n)$ for the range indicated.

(b) Using PC-DSP, generate the impulse response sequence $h(n)$. Use the menu selections *Sequences/Generate-sequence/Formula-entry-method*, and enter the analytical expression for $h(n)$

to be evaluated for $n = 0, \ldots, 8$. Compare the resulting sequence to your hand calculations in part (a).

Note: When computing samples of $h(n)$ using the formula-entry method, a division by zero will occur for $n = 4$. This can be prevented by adding a very small positive number to the index. In the analytical description of $h(n)$, replace each occurrence of n with $(n + 0.0001)$.

(c) Check to see if the impulse response obtained satisfies any of the linear-phase criteria established in Chapter 4. In general, does the design method used always guarantee linear phase?

Hint: Think about the symmetry properties of the DTFT discussed in Chapter 4. The ideal desired frequency response $H_d(\omega)$ was purely real. What kind of symmetry does $h_d(n)$ have? How does this symmetry carry over to $h(n)$?

(d) Compute the system function $H(\omega)$ of the designed filter using the discrete-time Fourier transform function accessed through menu selections *Transforms/DTFT*. Graph the magnitude and the phase of the system function. Is the phase response linear? Comment on the behavior of the magnitude response around the cutoff frequency.

Exercise 8.1.2

(a) Using the formula-entry method of PC-DSP as described in the previous exercise, design 21-, 31-, and 41-tap lowpass FIR filters with a cutoff frequency of $\omega_c = 0.4\pi$ radians. Tabulate and graph the magnitude and phase characteristics of each design. In the magnitude characteristics, how does the number of filter coefficients seem to relate to the passband ripple and to the sharpness of the passband edge? In each case, carefully examine the behavior of the magnitude response around the cutoff frequency. Do higher filter orders seem to reduce the oscillatory behavior of the magnitude response?

(b) While the DTFT graph is on the screen, pressing the L key causes the magnitude response to be graphed on a decibel scale. Recall that

$$|H(\omega)|_{dB} = 20 \log(|H(\omega)|).$$

Observe the magnitude characteristics in decibels for each filter designed. For each design, determine the decibel difference between the passband and the highest point of the stopband. Note that the highest point of the stopband is the peak of the first lobe outside the passband. This decibel difference is referred to as the *minimum stopband attenuation* and is a measure of how well the filter suppresses the frequencies outside the passband. Does the minimum stopband attenuation seem to increase with increasing number of filter coefficients?

As exercises (8.1.1) and (8.1.2) demonstrate, the magnitude response of an FIR filter designed using Eqs. (8.4), (8.5), and (8.7) exhibits oscillations around the cutoff frequency. The minimum stopband attenuation that can be obtained is about 20 dB regardless of the number of filter coefficients. This behavior is known as the *Gibbs phenomenon* and is due to the truncation of the ideal impulse response. To better understand the mechanism that leads to Gibbs phenomenon, we need to look at the spectral relationship between the ideal low-pass filter used as a starting point and the actual FIR filter designed. Substituting (8.5) into (8.7), we obtain

$$h(n) = \begin{cases} h_d(n - M), & 0 \leq n \leq 2M \\ 0, & \text{otherwise} \end{cases} \qquad (8.9)$$

The relationship between $h(n)$ and $h_d(n)$ given by (8.9) can be expressed in the compact form

Sec. 8.1 Fourier Series Design of FIR Filters

$$h(n) = h_d(n - M)w(n), \qquad (8.10)$$

where $w(n)$ is a *rectangular window* sequence defined as

$$w(n) = \begin{cases} 1, & 0 \le n \le N - 1 \\ 0, & \text{otherwise} \end{cases} \qquad (8.11)$$

and $N = 2M + 1$. Clearly, the impulse response of the designed filter is the product of the ideal low-pass filter impulse response with the rectangular window sequence. Thus, the frequency response of the designed filter must be the convolution of the corresponding spectra; that is,

$$\begin{aligned} H(\omega) &= [e^{-j\omega M} H_d(\omega)] * W(\omega) \\ &= \frac{1}{2\pi} \int_{-\pi}^{\pi} e^{-j\nu M} H_d(\nu) W(\omega - \nu)\, d\nu. \end{aligned} \qquad (8.12)$$

The frequency spectrum of the rectangular window sequence is

$$\frac{\sin(\omega N/2)}{\sin(\omega/2)} e^{-j\omega(N-1)/2}. \qquad (8.13)$$

The relationship between $H_d(\omega)$, $W(\omega)$, and $H(\omega)$ is illustrated in Fig. 8.4. This inherent convolution relationship is at the root of Gibbs oscillations. Now that we understand the cause of the problem, we are ready to arm ourselves with possible solutions.

Exercise 8.1.3

(a) Using PC-DSP, generate a 21-sample rectangular window sequence. Compute its frequency spectrum using the discrete-time Fourier transform function. Tabulate and graph the magnitude and the phase of the spectrum. Observe two things on the decibel magnitude plot: (1) the width of the main (center) lobe between two zero crossings, and (2) the decibel attenuation of the first side lobe on either side of the main lobe.

(b) Repeat the requirements of part (a) for 31-and 41-sample rectangular window sequences. How does the increased number of samples affect the main-lobe width and the side-lobe attenuation?

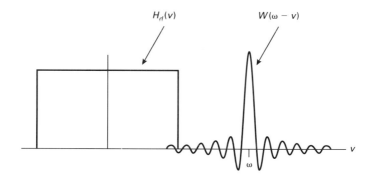

Figure 8.4 Graphical illustration of Gibbs phenomenon.

Gibbs oscillations are directly related to the oscillatory behavior in the frequency-domain characteristics of the rectangular window. The abrupt transition from unit amplitude to zero amplitude at the two edges of the rectangular window sequence corresponds to high-frequency components in its spectrum. The solution is to use an alternative window sequence $w(n)$ that tapers down smoothly and gradually at the edges, without generating high-frequency components. Another requirement is that the window sequence be symmetric so that the linear-phase property of the designed filter is preserved; that is,

$$w(n) = w(N - 1 - n), \quad \text{for all } n. \tag{8.14}$$

Thus, (8.10) is still valid as the design equation, and the only difference is that an alternative window sequence $w(n)$ is used instead of the rectangular window. A large variety of discrete window sequences is available. Analytical descriptions of some of them are given next.

Triangular (Bartlett) window

$$w(n) = \begin{cases} \dfrac{2n}{N-1}, & \text{if } 0 \le n \le \dfrac{N-1}{2} \\ 2 - \dfrac{2n}{N-1}, & \text{if } \dfrac{N-1}{2} < n \le N-1. \end{cases} \tag{8.15}$$

Hamming window

$$w(n) = 0.54 - 0.46 \cos\left(\frac{2\pi n}{N-1}\right), \quad 0 \le n \le N - 1. \tag{8.16}$$

Hanning (raised-cosine) window

$$w(n) = 0.5 - 0.5 \cos\left(\frac{2\pi n}{N-1}\right), \quad 0 \le n \le N - 1. \tag{8.17}$$

Blackman window

$$w(n) = 0.42 - 0.5 \cos\left(\frac{2\pi n}{N-1}\right) + 0.08 \cos\left(\frac{4\pi n}{N-1}\right), \\ 0 \le n \le N - 1. \tag{8.18}$$

Kaiser window

$$w(n) = \frac{I_0[\beta\sqrt{1 - (1 - 2n/(N-1))^2}]}{I_0(\beta)}, \quad 0 \le n \le N - 1. \tag{8.19}$$

The first four window sequences are shown in Fig. 8.5. Note that the envelope of each window sequence is shown as a continuous curve to make comparisons possible.

Two important features of all window functions are the width of the main lobe and the attenuation of the first side lobe in their spectra. Recall that, in the end, the frequency spectrum of the designed FIR filter is equal to the convolution of the ideal filter spectrum with the window spectrum. Thus, a narrow main lobe means a sharper cutoff in the

Sec. 8.1   Fourier Series Design of FIR Filters   139

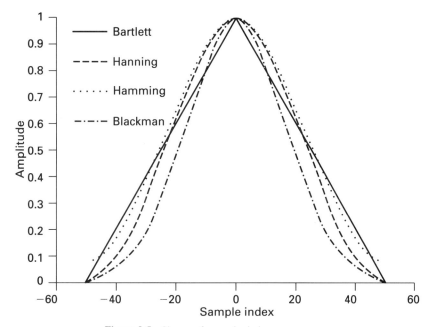

Figure 8.5 Shapes of several window sequences.

designed filter spectrum and is therefore desirable. On the other hand, increased side-lobe attenuation means better suppression in the stopband and is also desirable. Unfortunately, all window functions exhibit a trade-off between the width of the main lobe and the attenuation of the first side lobe. For a given value of N, better side-lobe attenuation is achieved at the expense of a wider main lobe and vice versa. The width of the main lobe can always be reduced by increasing the length of the window sequence. In contrast, the attenuation of the first side lobe depends only on the shape of the window, and not on its length.

In the case of the Kaiser window, $I_0(\beta)$ indicates the modified Bessel function of the first kind of order zero. The parameter β is used in adjusting the trade-off between the width of the main lobe and the decibel attenuation of the first side lobe. Shapes of Kaiser windows are shown in Fig. 8.6 for several values of the parameter β.

Exercise 8.1.4

(a) Using PC-DSP, generate a 21-sample triangular window sequence. Use the menu selections *Sequences/Generate-sequence/Discrete-window-functions*. When the data-entry window is displayed, select a triangular window and specify the number of samples. Compute the discrete-time Fourier transform of the generated window sequence. Tabulate and graph the transform. Record the width of the main lobe between its zero crossings and the decibel attenuation of the first side lobe.

(b) Repeat part (a) with a 31-sample triangular window. Compare the main-lobe width and the side-lobe attenuation to those obtained in part (a).

(c) Repeat part (a) with a 41-sample triangular window and compare to the previous cases.

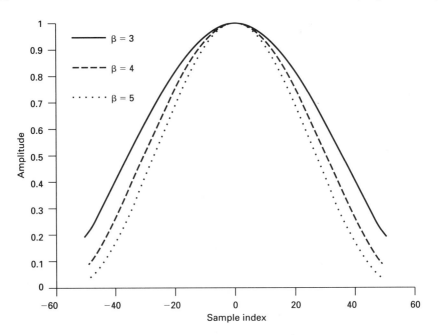

Figure 8.6 Shapes of Kaiser window sequences.

Exercise 8.1.5

Repeat the requirements of the previous exercise with (**a**) Hamming, (**b**) Hanning, and (**c**) Blackman windows. In each case, generate 21-, 31-, and 41-sample window sequences. Compute their spectra. Observe main-lobe width and side-lobe attenuation for each case, as was done in the previous exercise. Using your observations, complete the following two tables.

	Main-lobe Width (radians)		
	$N = 21$	$N = 31$	$N = 41$
Triangular	_____	_____	_____
Hamming	_____	_____	_____
Hanning	_____	_____	_____
Blackman	_____	_____	_____

	Side-lobe Attenuation (dB)		
	$N = 21$	$N = 31$	$N = 41$
Triangular	_____	_____	_____
Hamming	_____	_____	_____
Hanning	_____	_____	_____
Blackman	_____	_____	_____

Sec. 8.1   Fourier Series Design of FIR Filters   141

Exercise 8.1.6

In this exercise, we will redesign the low-pass filter of Exercises 8.1.1 and 8.1.2, this time using a Hamming window. The cutoff frequency of the desired ideal filter is $\omega_c = 0.4\pi$ radians.

(a) Using PC-DSP, generate the impulse response sequence $h(n)$ of a 21-tap FIR filter that approximates the desired ideal filter. Use the menu selections *Sequences/Generate-sequence/Formula-entry-method*, and enter the analytical expression for $h(n)$ that also contains the 21-sample Hamming window sequence as a factor. The impulse response we are seeking is (verify)

$$h(n) = \frac{\sin[0.4\pi(n-10)]}{\pi(n-10)} \left[0.54 - 0.46 \cos\left(\frac{2\pi n}{20}\right) \right], \quad n = 0, \ldots, 20.$$

After obtaining the impulse response sequence, compute its discrete-time Fourier transform. Graph the magnitude and the phase of the spectrum obtained. How does this design compare to the one obtained in Exercise 8.1.2 using a rectangular window? Comment on (1) the sharpness of the transition from the passband to the stopband and (2) the decibel stopband attenuation. Is the cutoff frequency obtained accurate?

(b) Repeat the design in part (a) using 31- and 41-tap filters. Does this seem to increase the accuracy of the cutoff frequency? As before, comment on the sharpness of the transition and the stopband attenuation.

Exercise 8.1.7

Consider again the 21-tap FIR low-pass filter designed in the previous exercise using a Hamming window. In this exercise, we will use that filter to process sinusoidal sequences.

(a) Generate samples of the sinusoidal sequence $x(n) = \cos(0.1\pi n)u(n)$ for the range of the sample index $0 \leq n \leq 99$. Processing this signal through the designed FIR filter can be accomplished by simply convolving the signals $x(n)$ and $h(n)$. Use menu selections *Operations/Processing-functions/Convolve-sequences* to obtain the output sequence. Plot the input and the output sequences in a superimposed fashion using the *Graphics* menu. How was the amplitude of the signal affected by the filter? Looking at the superimposed graph, can you estimate the delay caused by the filter? (*Hint:* Compare the locations of peaks of the two signals close to the right edge of the screen.) How does this estimate compare to the theoretical value for the delay?

(b) Repeat the requirements of part (a) for the input sequence $x(n) = \cos(0.6\pi n)u(n)$. Observe the steady-state portion of the output signal. Since the frequency of this sinusoid is in the stopband of the filter, it should be suppressed. Nevertheless, this behavior is observed after the first 20 samples of the response. Recall from Chapter 6 that it takes $N - 1$ sampling intervals for all states of an N-tap FIR filter to be filled.

Exercise 8.1.8

Even though the formula-entry method of generating sequences was used for the design of FIR filters in the preceding exercises, PC-DSP has a dedicated section for FIR filter design. Menu selections *Filters/FIR-filter-design/Fourier-series-method* lead to a data-entry window in which the desired filter type, number of filter coefficients, and the desired window function can be specified. The designed filter is saved into a binary data file with file name extension FLT. Once a filter data file is created in this fashion, menu selections *Filters/Analyze-filter* can be used to analyze it in terms of its magnitude and phase spectra, time and group delay characteristics, and impulse response.

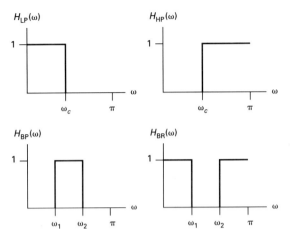

Figure 8.7 Ideal characteristics for frequency-selective filters.

(a) Using the approach described, design a 55-tap low-pass filter with the cutoff frequency $\omega_c = 0.15\pi$ radians. Use a triangular window. In specific filter design sections of PC-DSP, critical frequencies of the filter to be designed are entered using normalized values. Recall that $\omega = 2\pi F$, where F represents a percentage of the sampling rate. For the particular filter of this exercise, the normalized cutoff frequency is $F_c = 0.075$.

(b) Use the filter analysis section of PC-DSP to plot the magnitude and the phase of the designed filter. Note that a design report can also be viewed in the same section.

Other Filter Types

In the preceding discussion, we mainly concentrated on the design of low-pass filters. Nevertheless, the design technique described applies to other filter types as well. Figure 8.7 shows the ideal characteristics for four basic frequency-selective filter types. As we have established, an ideal low-pass filter has the frequency spectrum

$$H_{LP}^{(\omega_c)}(\omega) = \begin{cases} 1, & |\omega| < \omega_c \\ 0, & \omega_c < |\omega| < \pi, \end{cases} \quad (8.20)$$

where we have adopted a slight change in the notation. In the function $H_{LP}^{(\omega_c)}(\omega)$, the subscript indicates the type of the ideal filter (low pass), and the superscript indicates the cutoff frequency. The impulse response of an ideal low-pass filter was given in (8.4) and is repeated here with the same change of notation:

$$h_{LP}^{(\omega_c)} = \frac{\sin(\omega_c n)}{\pi n}. \quad (8.21)$$

Analytical forms of other filter characteristics can also be written rather easily. Consider an ideal high-pass filter with cutoff frequency ω_c.

$$H_{HP}^{(\omega_c)}(\omega) = \begin{cases} 0, & |\omega| < \omega_c \\ 1, & \omega_c < |\omega| < \pi. \end{cases} \quad (8.22)$$

Sec. 8.1   Fourier Series Design of FIR Filters   143

$H_{HP}^{(\omega_c)}$ can be expressed in terms of $H_{LP}^{(\omega_c)}$ as

$$H_{HP}^{(\omega_c)} = 1 - H_{LP}^{(\omega_c)}, \tag{8.23}$$

and the corresponding relationship between impulse responses of high and low-pass filters is

$$h_{HP}^{(\omega_c)}(n) = \delta(n) - h_{LP}^{(\omega_c)}(n). \tag{8.24}$$

Substitution of (8.21) into (8.24) yields

$$h_{HP}^{(\omega_c)} = \delta(n) - \frac{\sin(\omega_c n)}{\pi n}. \tag{8.25}$$

Exercise 8.1.9

Consider the 21-tap low-pass filter that was designed in Exercise 8.1.6 using a Hamming window. In this exercise, we will design a high-pass filter with the same cutoff frequency, $\omega_c = 0.4\pi$ radians. We will use the method described previously to obtain an analytical expression for the impulse response of this high-pass filter.

(a) Using PC-DSP, generate the impulse response sequence $h(n)$ of a 21-tap FIR filter that approximates the desired ideal high-pass filter. Use the formula-entry method, and enter the analytical expression for $h(n)$ that also contains the 21-sample Hamming window sequence as a factor. The impulse response we are seeking is (verify)

$$h(n) = \left[\delta(n-10) - \frac{\sin[0.4\pi(n-10)]}{\pi(n-10)}\right]\left[0.54 - 0.46\cos\left(\frac{2\pi n}{20}\right)\right], \quad n = 0, \ldots, 20.$$

(b) After obtaining the impulse response sequence, compute its discrete-time Fourier transform. Graph the magnitude and the phase of the spectrum obtained.

An ideal bandpass filter spectrum can be viewed as the difference of two ideal low-pass filter spectra:

$$H_{BP}^{(\omega_1,\omega_2)}(\omega) = H_{LP}^{(\omega_2)}(\omega) - H_{LP}^{(\omega_1)}(\omega). \tag{8.26}$$

Taking the inverse transform of each side of (8.26), we have

$$h_{BP}^{(\omega_1,\omega_2)}(n) = h_{LP}^{(\omega_2)}(n) - h_{LP}^{(\omega_1)}(n). \tag{8.27}$$

Finally, an ideal band-reject filter spectrum can be written in terms of the ideal bandpass filter spectrum:

$$H_{BR}^{(\omega_1,\omega_2)} = 1 - H_{BP}^{(\omega_1,\omega_2)} \tag{8.28}$$

with the corresponding time-domain relationship

$$h_{BR}^{(\omega_1,\omega_2)}(n) = \delta(n) - h_{BP}^{(\omega_1,\omega_2)}(n). \tag{8.29}$$

Exercise 8.1.10

(a) Using the Fourier series method with a Hamming window, a 21-tap FIR bandpass filter is to be designed with critical frequencies $\omega_1 = 0.3\pi$ and $\omega_2 = 0.5\pi$. Generate the impulse response sequence $h(n)$ of a 21-tap FIR filter that approximates the desired ideal bandpass filter. Use the formula-entry method of PC-DSP, and enter the analytical expression for $h(n)$ that also contains the 21-sample Hamming window sequence as a factor. The impulse response we are seeking is (verify)

$$h(n) = \left[\frac{\sin[0.5\pi(n-10)]}{\pi(n-10)} - \frac{\sin[0.3\pi(n-10)]}{\pi(n-10)}\right]\left[0.54 - 0.46\cos\left(\frac{2\pi n}{20}\right)\right],$$
$$n = 0, \ldots, 20.$$

(b) After obtaining the impulse response sequence, compute its discrete-time Fourier transform. Graph the magnitude and the phase of the spectrum obtained.

(c) Modify the analytical expression given in part (a) to obtain the impulse response of a bandreject filter with the same critical frequencies. Observe magnitude and phase characteristics.

8.2 FREQUENCY SAMPLING DESIGN

An alternative to the Fourier series design method is the *frequency sampling* technique. Let $H_d(\omega)$ be the desired frequency response to be approximated with an FIR filter. If this frequency response is sampled at N different frequencies ω_k in the range $0 \leq \omega \leq 2\pi$, we obtain

$$A_k = H_d(\omega_k), \quad k = 0, \ldots, N-1. \tag{8.30}$$

An N-tap FIR filter can be found such that its frequency response matches the desired frequency response *exactly* at the N frequencies selected; that is,

$$A_k = \sum_{n=0}^{N-1} h(n)e^{-j\omega_k n}, \quad k = 0, \ldots, N-1. \tag{8.31}$$

Equation (8.31) represents a system of N linear simultaneous equations that can be solved for the filter coefficients $h(n)$. It can also be written in matrix form as

$$\begin{bmatrix} A_0 \\ A_1 \\ \vdots \\ A_{N-1} \end{bmatrix} = \begin{bmatrix} 1 & e^{-j\omega_0} & e^{-j2\omega_0} & \cdots & e^{-j(N-1)\omega_0} \\ 1 & e^{-j\omega_1} & e^{-j2\omega_1} & \cdots & e^{-j(N-1)\omega_1} \\ \vdots & \vdots & \vdots & & \vdots \\ 1 & e^{-j\omega_{N-1}} & e^{-j2\omega_{N-1}} & \cdots & e^{-j(N-1)\omega_{N-1}} \end{bmatrix} \begin{bmatrix} h(0) \\ h(1) \\ \vdots \\ h(N-1) \end{bmatrix} \tag{8.32}$$

As a special case, if the frequencies ω_k are chosen to be equally spaced in the interval $0 \leq \omega \leq 2\pi$, then the inverse DFT equation can be used to obtain the solution in a much simpler way. With

$$\omega_k = \frac{2\pi k}{N}, \quad k = 0, \ldots, N-1,$$

we have

$$h(n) = \frac{1}{N}\sum_{n=0}^{N-1} A_k e^{j2\pi kn/N}, \quad n = 0, \ldots, N-1. \tag{8.33}$$

As established previously, the frequency response of the designed filter will exhibit an exact match to desired frequency response samples at N selected frequencies. It is also

Sec. 8.2 Frequency Sampling Design

interesting to see what the response looks like between these selected frequencies. The frequency response of the filter given by (8.33) is

$$H(\omega) = \sum_{n=-\infty}^{\infty} h(n) e^{-j\omega n}.$$

Summation limits can be changed to reflect the fact that $h(n) = 0$ outside the range $n = 0, \ldots, N-1$.

$$H(\omega) = \sum_{n=0}^{N-1} h(n) e^{-j\omega n}. \quad (8.34)$$

Substituting (8.33) into (8.34), we obtain

$$H(\omega) = \sum_{n=0}^{N-1} \left[\frac{1}{N} \sum_{k=0}^{N-1} A_k e^{j2\pi nk/N} \right] e^{-j\omega n}. \quad (8.35)$$

Summations in (8.35) can be rearranged to yield

$$H(\omega) = \frac{1}{N} \sum_{k=0}^{N-1} A_k \sum_{n=0}^{N-1} e^{-j(\omega - \omega_k) n}, \quad (8.36)$$

where we have also used $\omega_k = 2\pi k/N$. Finally, the inner summation can be expressed in closed form:

$$H(\omega) = \frac{1}{N} \sum_{k=0}^{N-1} A_k e^{-j\left(\frac{N-1}{2}\right)(\omega - \omega_k)} \frac{\sin[(N/2)(\omega - \omega_k)]}{\sin[(1/2)(\omega - \omega_k)]} \quad (8.37)$$

It can easily be verified from (8.37) that, at selected frequencies,

$$H(\omega_k) = A_k.$$

Between the selected frequencies, the frequency spectrum is obtained by interpolation. Note the similarity between (8.37) and (5.15), which was derived for the reconstruction of an analog signal from its samples.

Exercise 8.2.1

Using the frequency sampling design technique, a 27-tap FIR filter is to be designed to approximate the ideal low-pass spectrum with a cutoff frequency $\omega_c = 0.4\pi$. We will use the formula-entry method to generate samples of the desired frequency response.

(a) In the range $0 \leq \omega \leq 2\pi$, the desired ideal frequency response can be written using unit-step functions:

$$H_d(\omega) = u(\omega) - u(\omega - 0.4\pi) + u(\omega - 1.6\pi), \quad 0 \leq \omega \leq 2\pi.$$

Sampling $H_d(\omega)$ at N equally spaced frequencies, we obtain

$$\tilde{A}_k = H_d\left(\frac{2\pi k}{27}\right)$$

$$= u\left(\frac{2\pi k}{27}\right) - u\left(\frac{2\pi k}{27} - 0.4\pi\right) + u\left(\frac{2\pi k}{27} - 1.6\pi\right), \quad k = 0, \ldots, 27.$$

Using menu selections *Sequences/Generate-sequence/Formula-entry-method*, evaluate this expression for the range indicated. *Note*: In the formula-entry section of PC-DSP, the independent variable must be named n, so change all occurrences of k to n when entering the formula.

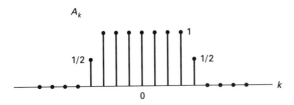

Figure 8.8 Using a transition band in frequency sampling design.

(b) Obtain the sequence $\tilde{h}(n)$ as the inverse discrete Fourier transform of $\tilde{A}_k$. Use the menu selections *Transforms/IDFT*, and compute the inverse transform with $N = 27$. Theoretically, the resulting sequence $\tilde{h}(n)$ should be real due to the symmetry of the transform; however, a complex sequence may be obtained with a very small imaginary part corresponding to machine round-off errors. If this is the case, remove the imaginary part using menu selections *Operations/Arithmetic-operations/Real-part*.

(c) Graph the sequence $\tilde{h}(n)$ obtained in part (b). Is it what you expected? You should find out that the sequence obtained does not satisfy any of the conditions necessary for linear phase. This is because our desired spectrum was not realistic. A causal linear-phase filter must have a delay of $(N - 1)/2$ samples. To remedy this problem, we need to add this delay to the impulse response. Since the inverse DFT was used in obtaining the impulse response, the delay must be added circularly. Use menu selections *Operations/Arithmetic-operations/Shift-sequence*, and circularly shift the sequence $\tilde{h}(n)$ by $(N - 1)/2 = 13$ samples; that is,

$$h(n) = \tilde{h}((n - 13))_{27}.$$

(d) Compute the frequency response of the filter using the DTFT function. Graph the magnitude and the phase characteristics. Is the phase characteristic linear now? Comment on the magnitude characteristic. How does it compare to those obtained in previous exercises using the Fourier series design method with a rectangular window? While the graph is on the screen, press *L* to display it in decibels and observe the minimum decibel stopband attenuation.

The magnitude characteristic of the low-pass filter designed in Exercise 8.2.1 should be very similar to that of a filter designed using the Fourier series design method with a rectangular window. It suffers from the same type of oscillatory behavior. This behavior is due to the fact that $\sin(x)/x$-type continuous interpolating functions are used for representing a discontinuity at the cutoff frequency. [See Eq. (8.37).] One possible method for improving the magnitude response of the filter and for increasing the minimum stopband attenuation is to use a transition band consisting of one or more frequency samples. Figure 8.8 illustrates the case where the last frequency sample in the passband is changed to 1/2.

Exercise 8.2.2

(a) Redesign the 27-tap low-pass filter of Exercise 8.2.1 using a one-sample transition band. Samples 5 and 22 of the desired frequency response A_k must be changed to 1/2. An easy way of obtaining the new set of frequency samples is to first generate the sequence

$$\epsilon(n) = 0.5\delta(n - 5) + 0.5\delta(n - 22), \quad n = 0, \ldots, 27$$

and then subtract it from the sequence of frequency-domain samples. Use menu selections *Operations/Arithmetic-operations/Subtract-sequences*. Afterward, use the inverse DFT and circular shift as in the previous exercise.

Sec. 8.2 Frequency Sampling Design

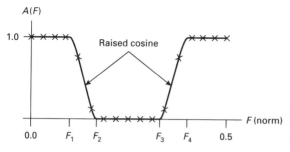

Figure 8.9 Use of a raised-cosine-shaped transition region in frequency-sampling design.

(b) Compute the frequency response of the filter using the DTFT function. Graph the magnitude and phase characteristics. Observe the behavior around the cutoff frequency. How does the decibel stopband attenuation compare to that obtained in Exercise 8.2.1?

Exercise 8.2.2 should demonstrate that changing amplitudes of the samples at passband edges reduce the oscillatory behavior of the magnitude response. A more general approach is to incorporate the transition-band behavior of the filter into the analytical form of the desired spectrum. For example, consider the desired magnitude response

$$H_d(\omega) = \begin{cases} 1, & 0 \le \omega \le \omega_1 \\ 0.5 + 0.5 \cos\left[\dfrac{\pi(\omega - \omega_1)}{(\omega_2 - \omega_1)}\right], & \omega_1 \le \omega \le \omega_2 \\ 0, & \omega_2 \le \omega \le \pi. \end{cases} \qquad (8.38)$$

This is depicted graphically in Fig. 8.9. A raised-cosine shape is used in the transition band to provide a smooth transition from the passband to the stopband. Note that $H_d(\omega)$ is continuous in magnitude and also in the first derivative.

Exercise 8.2.3

A 41-tap FIR low-pass filter is to be designed using the frequency-sampling technique with a raised-cosine-shaped transition region. The passband is required to be from 0 to 0.4π radians. The stopband should extend from 0.6π to π radians. In the range $0 \le \omega \le 2\pi$, the desired ideal frequency response can be written using unit-step functions (verify):

$$H_d(\omega) = u(\omega) - u(\omega - 0.4\pi) + u(\omega - 1.6\pi)$$
$$+ 0.5\left[1 + \cos\left(\dfrac{\omega - 0.4\pi}{0.2}\right)\right][u(\omega - 0.4\pi) - u(\omega - 0.6\pi)]$$
$$+ 0.5\left[1 + \cos\left(\dfrac{1.6\pi - \omega}{0.2}\right)\right][u(\omega - 1.4\pi) - u(\omega - 1.6\pi)].$$

(a) Using the formula-entry method of PC-DSP, evaluate samples of this desired response at 41 equally spaced frequencies; that is,

$$\tilde{A}_k = H_d\left(\dfrac{2\pi k}{41}\right), \qquad k = 0, \ldots, 40.$$

Note: In the formula-entry section of PC-DSP, the independent variable must be named n, so change all occurrences of k to n when entering the formula.

(b) Compute the impulse response of the corresponding FIR filter using inverse DFT and circular shift as in the previous two exercises. Compute the frequency response of the filter using the DTFT function. Graph the magnitude and phase characteristics.

Exercise 8.2.4

In the previous exercises of this section, we implemented the frequency-sampling design method using three steps: (1) compute samples of the desired magnitude behavior, (2) compute the inverse DFT, and (3) shift circularly to obtain the impulse response. PC-DSP has a built-in function for frequency-sampling design. It is accessed using menu selections *Filters/FIR-filter-design/Frequency-sampling-method* and produces FIR filter designs with raised-cosine-shaped transition regions.

(a) Using the frequency-sampling design function, redesign the filter specified in Exercise 8.2.3. In the data-entry window presented, select *multi-band filter* and specify the normalized critical frequencies as 0 and 0.2 for the first frequency band (the passband) and as 0.3 and 0.5 for the second frequency band (the stopband).

(b) Use the filter analysis section of PC-DSP to plot the magnitude and the phase of the designed filter. Note that a design report can also be viewed in the same section.

Exercise 8.2.5

The frequency-sampling design function of PC-DSP can also be used for designing frequency-sampling FIR filters without a transition band. The trick is to specify the band edge frequencies in such a way that no frequency samples fall into the transition band.

(a) Using menu selections *Filters/FIR-filter-design/Frequency-sampling-method*, design a 21-tap frequency sampling FIR low-pass filter with no transition band. The normalized passband cutoff frequency is required to be $F_1 = 0.2$. Recall that, for an N-tap filter, the frequency samples used in the design are spaced $1/N$ apart in terms of the normalized frequencies. With the proper choice of band edge frequencies, we can ensure that all frequency samples are ones and zeros, with no frequency samples taken in the raised-cosine-shaped transition band. Design a two-band filter with the normalized band edge frequencies 0, 0.2, 0.21, and 0.5. The last frequency sample in the passband is at the normalized $4/21 = 0.1904$, and the first frequency sample in the stopband is at $5/21 = 0.2380$. Thus, no samples are taken in the transition band $0.2 \leq F \leq 0.21$.

(b) Use the filter analysis section of PC-DSP to plot the magnitude and phase of the designed filter.

Exercise 8.2.6

(a) Redesign the FIR filter specified in Exercise 8.2.5 to have one transition sample with magnitude 0.5 at the normalized frequency $5/21$. To get a value of 0.5 at the normalized frequency $5/21$, you will need to ensure that this is the midpoint frequency of the raised-cosine transition band. Also, the edge frequencies of the transition band must be chosen close enough so that only one frequency sample falls within. Normalized frequencies $4.5/21$ and $5.5/21$ satisfy both requirements.

(b) Plot the magnitude and the phase of the designed filter. Compare the magnitude characteristic to that of the filter designed in Exercise 8.2.5.

Exercise 8.2.7

(a) Using the frequency-sampling design method, design a 21-tap notch filter to remove the normalized frequency $F = 6/21$. Note that this requires a three-band filter. The band edge frequencies should be selected in such a way that there is only one frequency sample in the rejection band (the stopband), and no frequency samples are taken in two raised-cosine transition bands. Following arrangement should work:

Passband: $F_1 = 0$, $F_2 = 5.25/21$
Stopband: $F_3 = 5.75/21$, $F_4 = 6.25/21$
Passband: $F_5 = 6.75/21$, $F_6 = 0.5$

(b) Generate 200 samples of the sequence

$$x(n) = \sin[2\pi(6/21)n]$$

and process with the filter designed. Use menu selections *Filters/Simulate-filter*. Comment on the result.

8.3 PARKS–MCCLELLAN ALGORITHM

The FIR filter design technique that is due to Parks and McClellan [20] is based on the idea of minimizing the maximum approximation error in a polynomial approximation to the desired filter magnitude response. The details of this design procedure are beyond the scope of this text and can be found in references [18], [19], and [20]. We will simply state the design criteria. Given a desired magnitude response $H_d(\omega)$, the approximation error can be written as

$$E(\omega) = W(\omega)[H(\omega) - H_d(\omega)], \quad (8.39)$$

where $H(\omega)$ is the system function of the desired linear-phase FIR filter expressed as a polynomial of $\cos(\omega)$ or $\sin(\omega)$, with coefficients yet to be determined. $W(\omega)$ is a weight function that is used for emphasizing a certain frequency band over others in the optimization process. The best approximation to the desired filter is found by minimizing

$$\max(|E(\omega)|)$$

over the set of filter coefficients. This results in an equiripple approximation to the desired magnitude characteristic.

Exercise 8.3.1

In this exercise, we will design a 25-tap equiripple FIR bandpass filter to approximate the desired magnitude response

$$H_d(\omega) = \begin{cases} 0, & 0 \le \omega \le 0.2\pi, & 0 \le F \le 0.1 \\ 1, & 0.25\pi \le \omega \le 0.45\pi, & 0.125 \le F \le 0.225 \\ 0, & 0.5\pi \le \omega \le \pi, & 0.25 \le F \le 0.5. \end{cases}$$

(a) In PC-DSP, use menu selections *Filters/FIR-filter-design/Parks-McClellan-algorithm*. Specify a multiband filter with three frequency bands. Enter the normalized band edge frequencies and the desired magnitude for each frequency band. Use uniform weighting.

(b) Analyze the resulting filter in terms of its magnitude behavior using the analysis functions in PC-DSP.

(c) Repeat parts (a) and (b) with a 45-tap FIR filter.

Exercise 8.3.2

(a) Using the Parks–McClellan algorithm, design a 45-tap FIR multiband filter to approximate the desired magnitude response

$$H_d(\omega) = \begin{cases} 0, & 0 \le \omega \le 0.3\pi, & 0 \le F \le 0.15 \\ 1, & 0.36\pi \le \omega \le 0.46\pi, & 0.18 \le F \le 0.23 \\ 0, & 0.52\pi \le \omega \le 0.6\pi, & 0.26 \le F \le 0.3 \\ 1, & 0.66\pi \le \omega \le 0.76\pi, & 0.33 \le F \le 0.38 \\ 0, & 0.8\pi \le \omega \le \pi, & 0.4 \le F \le 0.5. \end{cases}$$

Use uniform weighting for all frequency bands.

(b) Analyze the resulting filter in terms of its magnitude behavior using the analysis functions in PC-DSP.

(c) Generate 300 samples of the signal

$$x(n) = 3\sin(0.2\pi n) + 4\cos(0.4\pi n) - 5\cos(0.56\pi n) + 8\sin(0.72\pi n)$$

for $n = 0, \ldots, 299$. Compute and graph the DTFT of this signal.

(d) Process the signal obtained in part (c) with the designed multiband filter using menu selections *Filters/Simulate-filter*. Compute and graph the DTFT of the output signal. Comment.

Exercise 8.3.3

(a) Using the Parks–McClellan algorithm, design a 25-tap FIR differentiator with unit slope. Choose the frequency band of interest between normalized frequencies 0.05 and 0.45. Only one frequency band is specified for a differentiator.

(b) Plot the impulse response of the resulting filter. Is it what you would intuitively expect? Think of what the impulse response of an ideal differentiator would look like.

(c) Generate 300 samples of the sinusoid

$$x(n) = 3\sin(0.2n), \quad n = 0, \ldots, 299,$$

and process with the FIR differentiator designed in part (a). Compare the output sequence to the theoretical derivative of $x(n)$. *Hint*: Don't forget to take the 12-sample delay of the FIR filter into account.

(d) Repeat part (c) using a periodic pulse train instead of a sinusoid. Generate 10 periods of the pulse train, with each pulse consisting of 15 unit-amplitude samples and 10 zero-amplitude samples.

Exercise 8.3.4

(a) Using the Parks–McClellan algorithm, design a 33-tap Hilbert transform filter with optimum response in the normalized frequency range 0.05 to 0.45. Note that there is only one frequency band for a Hilbert transform filter. Also, the first and last edge frequencies are not required to be 0.0 and 0.5 as in the multiband filter case.

(b) A length-300 sequence is available in the data file EX8-3-4.SEQ that is on PC-DSP distribution disks. Compute the DTFT of this sequence. Graph the magnitude and the phase of the transform.

(c) Process this sequence with the Hilbert transform filter designed. Compute and graph the DTFT of the output sequence. How does the phase of the output signal compare to that of the input signal?

A

PC-DSP Reference

The disk(s) accompanying this textbook contain the *student edition* 2.00 of PC-DSP written for IBM compatible personal computers. PC-DSP (Personal Computer Digital Signal Processing) is an interactive, menu-driven software package used for the analysis, design, and implementation of discrete-time signals and systems. It can be considered a special-purpose calculator geared toward performing digital signal-processing operations such as waveform synthesis and manipulation, Fourier transforms, convolution, correlation, filter design, analysis and implementation, power spectrum analysis, and graphics. You will see that it is simple and straightforward to use, and yet powerful and extensible enough to be used by practicing engineers as well as students of digital signal processing. At this point, one word of caution is in order: No computer program eliminates the need for learning the underlying theory of the particular subject matter at hand. PC-DSP is intended to be used as a mathematical tool to enhance the understanding of digital signal-processing concepts and to assist in calculations that would be tedious to carry out manually. It is always up to the student, however, to interpret the results correctly and to use them in a meaningful way.

This appendix will serve as a brief reference guide for the program. Basic features of PC-DSP will be presented, and a brief overview of its capabilities will be given. Section A.1 contains hardware requirements and a brief introduction to the user-interface shell of the program. Section A.2 contains detailed descriptions of specific functions of the program. Most of this information is also available in the on-line help system. Section A.3 details PC-DSP data-file specifications.

A.1 BASIC CONCEPTS

To use PC-DSP student edition 2.00, you will need:

- An IBM PC or 100% compatible computer operating under the Microsoft Disk Operating System (MS-DOS) version 3.0 or higher.
- At least 512 K-bytes of RAM (random-access memory) installed in the system. 640 K-bytes is recommended if user-written programs are to be interfaced to PC-DSP for advanced applications.
- Graphics display capability (CGA, EGA, VGA graphics adapter).
- A fixed disk drive is strongly recommended. Although it is possible to configure the program to run on a floppy-disk-based system (at least 1-MB total storage capacity required), this results in significantly slower operation due to the use of overlays and data files.
- A math coprocessor is optional. If it is present in the system, PC-DSP will automatically detect it and use it to speed up floating-point calculations.
- The use of a mouse, although not required, greatly simplifies navigating through the menu structure of the program.
- A printer is not required, but is recommended for obtaining hard copies of tabulations and graphics. To obtain printer graphics, you will also need the proper graphics driver program for your printer (GRAPHICS.COM or similar).
- For proper operation of PC-DSP, it is recommended that the following two lines be placed in the CONFIG.SYS file of the disk operating system:

```
Files = 20
Buffers = 20
```

The PC-DSP program code is contained in a number of files with the DOS file name extension EXE. The file PCDSP.EXE is the shell that implements the components used for the display and for interaction with the user. These components include the menu structure, the status bar, dialog boxes for data entry and for tabulation of results, a progress monitor to provide information about the computations performed, an editor for developing macros, and on-screen graphics. Other files with the EXE extension are overlays. They contain code for the numerical algorithms.

For instructions on how to install PC-DSP on a computer, refer to Chapter 1. Once installed, PC-DSP can be started by switching to the program directory and typing the program's name without the DOS file extension. Assuming that the program is installed on fixed-disk drive C, in the subdirectory named DSP, the following two lines would be needed:

```
cd c:\dsp        <ENTER>
pcdsp            <ENTER>
```

Note that the word ⟨ENTER⟩ is not meant to be typed in. It just indicates that the ⟨ENTER⟩ key (labeled ⟨RETURN⟩ on some keyboards) should be pressed after typing each line. It

Sec. A.1 Basic Concepts

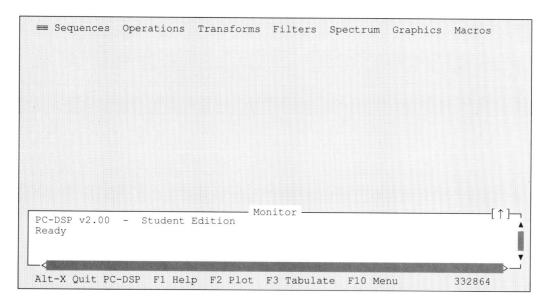

Figure A.1 PC-DSP opening screen.

is also possible to start the program from a remote directory if the PATH statement in the DOS start-up file AUTOEXEC.BAT contains a reference to the subdirectory where PC-DSP files are installed.

When PC-DSP is started, the opening screen shown in Fig. A.1 is displayed. Pressing the ⟨ENTER⟩ key or the space bar removes the information window in the middle of the screen. The top row of the screen is occupied by a menu bar that lists the main sections of PC-DSP. The status line is the bottom row of the screen. It contains information about special keys for obtaining on-line help, graphing or tabulating the last sequence computed, gaining access to the menu structure, and exiting the program. Also, the available memory is displayed at the lower-right corner of the screen. Right above the status bar, there is a progress monitor. This is a TTY-type scrolling window that displays information about the computations performed. If desired, it can be expanded to cover the entire area between the menu bar and the status bar. Its contents may be viewed using the scroll bars to the right of it.

When one of the items is selected in the top menu bar, a pull-down menu is presented with further selections in that category. Some selections in pull-down menus lead to dialog windows where parameter values can be entered, a desired action can be specified, or choices can be made as required by the selected function. Some other selections lead to additional pull-down menus.

There are a number of methods for making menu selections depending on the input device used. If a mouse is available, positioning the mouse cursor on the desired item and pressing the left mouse button is all that is necessary. In the absence of a mouse, the F10 key can be used to gain access to the menu system. Directional arrow keys can then be used for horizontal and vertical movement within the menu structure, and the ⟨ENTER⟩ key finalizes the highlighted menu selection. Alternatively, a top menu bar item can be selected by pressing the *Alt* key and the highlighted letter of the desired item simultaneously; for example, *Alt-O* selects the item entitled *Operations* and displays the corresponding pull-down menu. Once the pull-down menu is displayed, vertical selections are made by pressing the highlighted letter of the desired item without the *Alt* key.

The items on the status line can also be selected by either using the mouse or pressing the keys indicated. *Alt-X* terminates the program after the user confirms this in a dialog window presented. The *F1* key provides access to the on-line help system, which has its own internal menu structure. The *F2* key graphs the sequence generated as a result of the most recently used function, while the *F3* key numerically tabulates the same. The *F10* key provides access to the top menu bar. This is especially handy if you don't have a mouse.

In PC-DSP, the basic data structure is a *sequence* that consists of a number of samples that are either real or complex valued. Sequences are identified with alphanumeric names that may be up to eight characters long. The first character in the name of a sequence is required to be nonnumeric. Also, sequence names cannot contain any punctuation characters. Most PC-DSP operators operate on sequences, and they also generate sequences as their results. PC-DSP sequences are stored in the designated data directory as binary data files. The name of the binary file is the sequence name followed by the extension SEQ. In the student edition, any particular sequence can be up to 1024 samples long.

In addition to a sequence, several other data structures are also used. Analog and digital filters are stored in binary files with the extension FLT. Analog signals are stored in files with the extension SIG. (The only instance of this in the student version is the DTFT result stored as an analog signal.) Sequences saved as ASCII files are given the extension TXT.

The user's interaction with the program takes place in dialog windows where necessary parameters are entered. Figure A.2 shows the dialog window that is presented when the *Shift-sequence* function is selected using menu choices *Operations/Arithmetic-operations/Shift-sequence*. As shown, a dialog window typically consists of several fields and buttons. Any field can be made current by pressing the left mouse button while the mouse cursor is on that field. Also, the tab key can be used to cycle through the fields on a dialog window. Within a multiline field (for example, list boxes, clusters of check boxes, and option buttons), directional keys can be used to move the highlight. Most dialog windows have two buttons, one labeled *OK* and one labeled *Cancel*. As might be guessed, the *OK* button is used for carrying out the specific function requested through that dialog window. Conversely, the *Cancel* button allows the user to exit the dialog window without completing its action. Buttons on a dialog window can be pressed either with the mouse or by making the button current and pressing the ‹ENTER› key. Keyboard shortcuts can also be used, for example, *Alt-O* for the *OK* button and *Alt-C* for the *Cancel* button.

An input field is where alphanumeric information can be entered. The example *Shift-sequence* dialog window contains input fields for the names of the sequences involved, as well as the number of samples for the shift. If an input field has a little button with a down-arrow character in it, then possible entries for that field (such as names of existing sequences) can be obtained by clicking on that button with the mouse or by pressing the down-arrow key on the keyboard while the input field has the focus.

Sec. A.1   Basic Concepts

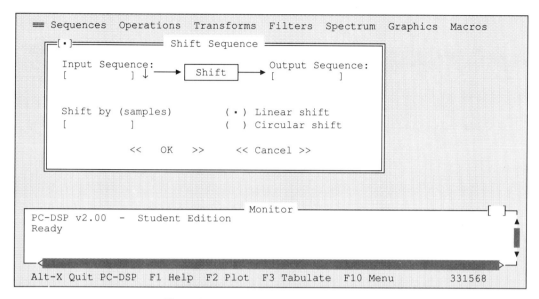

Figure A.2 Dialog window for the *Shift-sequence* function.

In the given dialog window example, the choices *Linear shift* and *Circular shift* represent a cluster of option buttons. One of these can be selected by making the cluster the current field and then using the up-and down-arrow keys.

Some input fields expect numerical data. It is important that the value entered be in the correct type that the program expects. In the example dialog window of Fig. A.2, the input field that accepts the number of shifted samples is an integer field, and entering a fractional (floating-point) value would result in an error.

In input fields that accept complex numbers, a complex number is entered by its real and imaginary parts separated with a comma. For example, the number $-1.2 + j3.2$ is entered as

$$-1.2,3.2$$

If a sample is purely imaginary, then it is sufficient just to enter its imaginary part preceded by a comma. For example, the number $j5$ can be entered as

$$0,5$$

or equivalently

$$,5$$

A.2 FUNCTION REFERENCE

In this section, details of the specific functions in PC-DSP will be given in reference format. The section is divided into subsections that correspond to the items on the top menu bar. Each subsection is further divided based on the items in corresponding pull-down menus.

A.2.1 System Menu

The system menu is accessed by clicking on the symbol ≡, which is on the left side of the menu bar. Without a mouse, the F10 key is used to gain access to the menu bar, followed by left and right arrow keys to select the system menu.

About PC-DSP. Displays copyright information about PC-DSP as well as the current revision number and date.

Temporary Exit to DOS. This function is used for temporarily returning control to the operating system so that system-related tasks can be performed; that is, you can get a directory listing, copy or delete files, or run another program. To get back to PC-DSP, type EXIT at the operating system prompt.

Quit PC-DSP. This terminates the program and returns control to the operating system. To prevent accidental exits, the user is given a chance to change his or her mind before the program terminates. Selecting this menu item is equivalent to pressing the key combination *Alt-X* or selecting the corresponding text area on the status line. In contrast with the previous version of PC-DSP, terminating the program does not result in the loss of the sequences created, since all data are now stored on disk. When the program is run at a later time, all sequence data from the previous session will be available unless the data files are manually deleted.

A.2.1.1 Configuration Options Submenu
This submenu contains selections that are used for changing the configuration settings (that is, locations of program and data files, screen size and colors, mouse parameters, and such) of PC-DSP.

Directories. Presents a dialog window to allow the user to specify subdirectories where certain types of files are to be kept. By default, all files generated during the operation of the program are written to and read from the current directory. This may be undesirable since a large number of data files may be generated during a typical session. It is usually easier to maintain and update data files if different file types are kept in different subdirectories.

The *data directory* is where data files with extensions SEQ, SIG, FLT, and TXT are stored. (These file-name extensions correspond to data files for discrete sequences, continuous signals, linear filters, and ASCII text, respectively.)

The *log files directory* is where files with the extension LOG are stored. These files are used to record user responses to certain items in each dialog window. In subsequent uses of a dialog window, PC-DSP looks for its log file. If one exists, its contents are used to determine default responses for that dialog window.

The *temporary directory* is used for temporary data files. Some PC-DSP routines generate temporary data files during certain operations. These files are automatically deleted when they are no longer needed. If no temporary directory is specified, the current directory is used. In some cases, performance can be improved by using a directory other than the current directory. This is especially true if a RAM disk is available and if it can be used for temporary data storage. (About 100 K-bytes should be sufficient for most purposes.)

The *macro files directory* is used to store PC-DSP macros, the extensions to PC-DSP. Several macro files are provided on the distribution disks and are referred to in some of the exercises. Macros are binary program files that are created using the PC-DSP macro compiler.

Colors. Used for modifying the screen colors for the menu system, status line, dialog window elements, editors, and the like.

Display Mode. Used for specifying the desired display mode. Available choices are color display, black and white display on a color monitor, and monochrome display. The number of screen rows can be specified as 25, 43, or 50 (the last two are only available for VGA graphics adapters).

Mouse Options. Allows the user to adjust mouse speed parameters. Double-click delay is the maximum time that can elapse between two successive mouse clicks in a double clicking of the mouse. *Repeat delay* is the time delay between the pressing of a mouse button and the start of repetition of mouse events. It is also possible to reverse the roles of left and right mouse buttons by checking the appropriate check box.

Save Options. Allows the user to save configuration information into a file with the name PCDSP.CFG so that the program is automatically set for that configuration in subsequent uses. The configuration file is created in the subdirectory that contains the executable files. If it is not present, the default configuration is assumed.

A.2.1.2 Editor Submenu

PC-DSP has a built-in general-purpose text editor that can be used for viewing and editing filter design reports or for developing PC-DSP macros. Another use of the text editor is for creating an ASCII file containing sample values of a sequence. This file can later be imported into PC-DSP as a sequence using the menu selections *Sequences/Import-ASCII-data-file*.

The editor window can be sized, zoomed, and moved. A block can be marked using the shift key and one of the directional arrow keys simultaneously. A marked block can be cut (that is, copied to the clipboard) with the *delete* key. The contents of the clipboard can be inserted into the document at the current cursor position using the *insert* key.

Several editor windows can be on the screen at one time. To switch from one window to another, either click on the new window with the left mouse button or choose the option *Next window* from the *Editor* submenu. An editor window can be resized by dragging its lower-right corner with the mouse. It can be moved on the screen by dragging its title bar. Alternatively, the appropriate options of the *Editor* submenu can be used to achieve the same without a mouse.

An editor window can be closed by pressing the close button at the upper-left corner or by selecting the *Close* option from the *Editor* submenu. If its contents have been modified, you will be asked if you would like to save the contents into a file before closing.

Create New File. Opens a blank text editor window ready to accept text. Before exiting the editor, the user is given the option of saving the text typed into the editor.

Open Existing File. Opens a new editor window and loads an existing text file into it. First, a file selection dialog window is presented. An existing ASCII text file can be selected either by typing its name or by navigating through the directory structure.

Size/Move Window. Used for resizing or moving the current editor window. The directional arrow keys can be used for moving the editor window on the screen. The size of the editor window can be changed by pressing the *Shift* key simultaneously with an arrow key. The window must be made current before it can be sized or moved. This selection also allows sizing and moving the sequence tabulation windows, as well as the progress monitor. (See Section A.2.2.)

Zoom Window. Toggles the size of the editor window between its default and maximum (full screen) sizes. The window must be made current before it can be zoomed. This selection can also be used for zooming the information monitor.

Next Window. If more than one editor window is on the screen, this command gives the focus to the next window in the list of editor windows. Additionally, the information monitor and any sequence tabulation windows that are on the screen are considered to be part of this list and can be made current using this function.

Previous Window. If more than one editor window is on the screen, this command gives the focus to the previous editor window in the list of editor windows. Additionally, the information monitor and any sequence tabulation windows that are on the screen are considered to be part of this list and can be made current using this function.

Close Window. Closes the current editor window. If its contents have been modified, a dialog window will appear asking if the contents should be saved into a file.

A.2.2 Sequences Menu

This submenu contains functions that are used for creating new sequences as well as viewing existing sequences, deleting sequences, and performing conversions between data files in different formats.

List Sequences. Provides a list of discrete sequences stored in the current data directory. (This directory is specified using the *Directories* option of the *Configuration Options* submenu. See Section A.2.1.1.) Type, size, and starting index of each sequence are also listed. The *current sequence* is also shown. (The current sequence is usually the result of the most recent operation. Recall that this is the sequence that is graphed with the *F2* key and tabulated with the *F3* key.) Any sequence in the list can be made current by double clicking on its name. Without a mouse, the direction keys and the ‹*ENTER*› key can be used instead.

Sec. A.2 Function Reference

Tabulate Sequence. Allows on-screen tabulation of a specified sequence. A new window is created with sample values of the sequence. More than one sequence tabulation can be on the screen at one time. Tabulation windows can be sized, moved, or closed using the corresponding menu selections in the *Editor* submenu. (See Section A.2.1.2.)

Plot Sequence. Generates an on-screen plot of a specified sequence. If the sequence is complex, its real and imaginary components will be shown separately. While the plot is on the screen, some keys have special meanings:

- *C:*   Continuous plot
- *D:*   Discrete plot
- *G:*   Toggle grid on and off
- *M:*   Toggle between Cartesian and polar representations

Delete Sequence. This menu choice is used for deleting a specified discrete-time sequence from the disk. The user is asked to confirm this decision before the sequence is deleted.

Import ASCII Data File. Reads an ASCII format text file containing sample values of a sequence and converts it into the binary data file format of PC-DSP. This function is typically used to import data files produced by other programs into PC-DSP. It can also be used in conjunction with the built-in text editor of PC-DSP for generating PC-DSP sequences by typing numbers into a text file and importing the file.

Upon selecting this function, the user is presented with a file-selection dialog window so that an existing ASCII data file can be selected. The file to be imported should contain numbers separated by commas or line breaks. No control characters are allowed. It is possible to import an ASCII file into PC-DSP as either a real-valued or complex-valued sequence. If it is imported as a real-valued sequence, the numbers in the file are read as sample values. If the file is imported as a complex-valued sequence, then the numbers in the file are read in pairs. In this case, the first number in each pair is used as the real part of the corresponding sample, and the second number becomes the imaginary part. In either case, the starting index is specified by the user.

Save as ASCII Data File. Writes sample values of a PC-DSP sequence into an ASCII format text file. This function is typically used to export PC-DSP sequences into other programs.

If the sequence under consideration is real valued, then its samples are written into the ASCII text file with one number on each line. For a complex sequence, a pair of numbers is written on each line representing real and imaginary parts of each sample.

Import MATLAB Data File. Reads a data file written by PC-Matlab™ and converts its contents into the binary data file format of PC-DSP. Upon selecting this function, a dialog window is presented with fields for the name of a data file and the name of the variable to be imported from that data file. Pressing the *Files* button generates a file-selection dialog window so that an existing data file can be selected for import. The file to be imported must have the file-name extension MAT. Once a data file is selected, pressing the *Contents* button produces a list of variables in that file. Since a PC-Matlab™

data file may contain more than vector or matrix, the user needs to select which vector is to be imported. Note that only vectors (not matrices) can be imported. PC-Matlab™ vectors always start at index 1, but when they are imported into PC-DSP, the starting index is specified by the user.

Save as MATLAB Data File. This function writes sample values of a PC-DSP sequence into a file in a binary format compatible with PC-Matlab™. The output file is given the file-name extension MAT.

A.2.2.1 Generate Sequence Submenu

This submenu contains functions for creating new sequences. A new PC-DSP sequence can be created in a number of ways. It can be described by means of a formula or by means of its statistical characteristics. Its sample values can be specified one by one. Alternatively, one of the predefined mathematical definitions can be used with key parameters specified by the user.

Read from Keyboard. Allows the creation of a new sequence by typing sample amplitudes one by one. Upon selecting this function, the user is presented with a dialog window that contains six input lines. Sample amplitudes should be typed into these lines, separated from each other by semicolons. Real and imaginary parts of a complex sample are separated by commas. Up to 60 real- or complex-valued samples can be entered into a sequence. Note that it is not necessary to fill all six input lines.

An alternative method of generating a sequence sample by sample is to use the built-in text editor. Type one sample amplitude per line, and save the resulting file with an extension TXT. Use the Import ASCII file function to convert it to a PC-DSP sequence. This function is accessed through menu selections *Sequences/Import-ASCII-file*.

Formula-entry Method. This function facilitates the creation of a new sequence from its mathematical description. A sequence is generated by evaluating a mathematical expression in a specified range of an integer index. For example, consider a sequence in the form

$$x(n) = \sin(0.6 \pi n)[u(n) - u(n - 100)].$$

To generate samples of this sequence, enter its formula into the formula field as

```
sin(0.6*pi*n)*(step(n) - step (n - 100))
```

Note that only the right side of the mathematical expression needs to be entered. The index variable must be n. The formula entered must be syntactically correct in terms of parentheses matching and function arguments. The precedence of arithmetic operators is the same as in most programming languages (for example, C, Pascal, or FORTRAN), and parentheses can be used to change the order of operators. The following arithmetic operators are supported:

- \+ (addition)
- \- (subtraction)
- * (multiplication)

Sec. A.2   Function Reference   161

    /   (division)

    ^   (raise to power)

Complex numbers can be entered by using braces. Inside braces, real and imaginary parts of a number are separated by a comma. For example, the expression

$$\mathtt{exp(\{1,2\}*n)*step(n)}$$

represents the sequence $x(n) = e^{(1 + j2)n}u(n)$. Real and imaginary parts of a complex number can be mathematical expressions themselves. For example,

$$\mathtt{\{cos(0.2*pi*n),sin(0.2*pi*n)\}}$$

is valid. The intrinsic functions supported are:

sin(x)	cos(x)	tan(x)	atan(x)	atan2(x, y)
sinh(x)	cosh(x)	tanh(x)	exp(x)	ln(x)
log(x)	abs(x)	real(x)	imag(x)	round(x)
trunc(x)	sqr(x)	sqrt(x)	imp(n)	step(n)
sign(x)	gau(m, v)	uni(l, u)		

sin(x):	Sine function: the argument x should be in radians.
cos(x):	Cosine function: the argument x should be in radians.
tan(x):	Tangent function: the argument x should be in radians.
atan(x):	Arc tangent function: the return value is in radians and is in the range $(-\pi/2, \pi/2)$.
atan2(x, y):	Same as atan(y/x). The return value is radians and is in the range $(-\pi, \pi)$.
sinh(x):	Hyperbolic sine function.
cosh(x):	Hyperbolic cosine function.
tanh(x):	Hyperbolic tangent function.
exp(x):	Exponential function.
ln(x):	Natural logarithm of x: the argument x must not be equal to zero.
log(x):	Base 10 logarithm of x: the argument x must not be equal to zero.
abs(x):	Absolute value of x: if x is complex, abs(x) returns the magnitude.
real(x):	Real part of x.
imag(x):	Imaginary part of x.
conj(x):	Complex conjugate of x.
round(x):	Returns the integer closest to x.
trunc(x):	Returns the largest integer less than x.
sqr(x):	Square of x.
sqrt(x):	Square root of x.
imp(n):	Unit-impulse sequence value at index n.

step(*n*): Unit-step sequence value at index *n*.
sign(*x*): Signum function: returns 1.0 if $x \geq 0$; returns 0.0 otherwise.
gau(*m*, *v*): Returns a random number from a Gaussian distribution with mean *m* and variance *v*.
uni(*l*, *u*): Returns a random number from a uniform distribution with lower and upper limits *l* and *u*, respectively.

Discrete Window Functions. Generates a discrete window sequence of a specified kind with a specified number of samples. The user enters the type of window function to be generated and the number of samples desired. Generates a real-valued sequence and starts at index $n = 0$. The following window functions are supported:

Rectangular window

$$w(n) = u(n) - u(n - N)$$

Triangular (Bartlett) window

$$w(n) = \begin{cases} \dfrac{2n}{N-1}, & \text{if } 0 \leq n \leq \dfrac{N-1}{2} \\ 2 - \dfrac{2n}{N-1}, & \text{if } \dfrac{N-1}{2} < n \leq N-1 \end{cases}$$

Hamming window:

$$w(n) = 0.54 - 0.46 \cos\left(\frac{2\pi n}{N-1}\right), \quad 0 \leq n \leq N - 1$$

Hanning (raised-cosine) window

$$w(n) = 0.5 - 0.5 \cos\left(\frac{2\pi n}{N-1}\right), \quad 0 \leq n \leq N - 1$$

Blackman window

$$w(n) = 0.42 - 0.5 \cos\left(\frac{2\pi n}{N-1}\right) + 0.08 \cos\left(\frac{4\pi n}{N-1}\right), \quad 0 \leq n \leq N - 1$$

Kaiser window

$$w(n) = \frac{I_0[\beta\sqrt{1 - (1 - 2n/(N-1))^2}]}{I_0(\beta)}, \quad 0 \leq n \leq N - 1$$

In the case of the Kaiser window, $I_0()$ indicates the modified Bessel function of the first kind of order zero. The parameter β is used in adjusting the main-lobe width and side-lobe attenuation. For theoretical details of window functions, refer to the section on Fourier series design of FIR filters.

Random Sequences. This menu selection is used for generating pseudorandom sequences. Even though they are generated using deterministic methods, they mimic the

Sec. A.2 Function Reference

characteristics of random signals. Approximations to three probability density functions are provided: uniform, Gaussian, and exponential. The user specifies the distribution desired, the number of samples required, and any relevant parameters of the distribution. The random sequence generated is real valued and starts at index $n = 0$.

Uniform random numbers are generated using the multiplicative congruental method, followed by shuffling. Initially, a random number between 0.0 and 1.0 is generated and then scaled into the desired range.

Random numbers with Gaussian distribution are obtained from uniform deviates using the Box–Muller method. Given two random numbers x_1 and x_2 from a uniform distribution between 0.0 and 1.0, two new numbers y_1 and y_2 are computed as

$$y_1 = \sqrt{-2 \ln(x_1)} \cos(2\pi x_2),$$
$$y_2 = \sqrt{-2 \ln(x_1)} \sin(2\pi x_2).$$

Using the concept of random variable transformations, it can be shown that y_1 and y_2 represent numbers from a Gaussian distribution with zero mean and unit variance. For other values of the mean and the variance, they can easily be scaled.

Random numbers with exponential distribution are also generated starting with uniform deviates. For exponential distribution, the transformation used is

$$y = -\ln(x).$$

Waveform Generator. This menu selection allows some simple signals to be generated. The user specifies the signal type desired, the number of samples to be generated, and any relevant parameters. The signal generated can be real or complex valued depending on the parameter values entered, and it starts at index $n = 0$. The following signal types are supported:

Impulse sequence: $x(n) = A\delta(n)$.
Step sequence: $x(n) = Au(n)$.
Ramp sequence: $x(n) = (A + Sn)u(n)$.
Sine: $x(n) = A \sin(Bn + C)u(n)$.
Cosine: $x(n) = A \cos(Bn + C)u(n)$.
Exponential: $x(n) = Ae^{(Bn + C)}u(n)$.
Right-sided exponential: $x(n) = AB^n u(n)$.

A.2.2.2 Sequence Editing Submenu

This submenu contains functions that are used for cut-and-paste type of editing of discrete sequences.

Copy Sequence. This function creates a new sequence by copying a user-specified range of samples of the input sequence. For a given input sequence $x(n)$ and a copy range of (N_1, N_2), the output sequence $y(n)$ is computed as

$$y(n) = \begin{cases} x(n), & \text{if } N_1 \leq n \leq N_2 \\ 0, & \text{otherwise.} \end{cases}$$

The input sequence $x(n)$ does not change after the operation.

Make Complex. This function is used for creating a complex sequence by combining two real-valued input sequences. Specifically, using the input sequences $x_1(n)$ and $x_2(n)$, the output sequence $y(n)$ is computed as

$$y(n) = x_1(n) + jx_2(n).$$

If the input sequences $x_1(n)$ and $x_2(n)$ are not real, this equation is still used; however, a warning message is printed on the progress monitor. If the input sequences $x_1(n)$ and $x_2(n)$ are of unequal lengths or if their starting index values are not equal, zero padding is used. The input sequences $x_1(n)$ and $x_2(n)$ do not change after the operation.

Make Periodic. This function creates a periodic sequence by repeating all or part of an input sequence. Given an input sequence $x(n)$ that starts at index $n = 0$ and a period length N, the output sequence $y(n)$ is computed as

$$y(n) = x((n))_N,$$

where double parentheses indicate modulo-N operation. Initial and final index values for the output sequence are also specified by the user.

Copy Segment from Sequence. This function is used for copying a user-specified range of samples of the input sequence into a new sequence with the name $CLPBRD$. For a given input sequence $x(n)$ and a copy range of (N_1, N_2), the output sequence $y(n)$ is computed as

$$y(n) = \begin{cases} x(n), & \text{if } N_1 \leq n \leq N_2 \\ 0, & \text{otherwise} \end{cases}$$

and stored in the data file $CLPBRD$.SEQ. Note that this function is identical to the *Copy-sequence* function described previously, with the only difference being the name of the output sequence. The segment copied in this fashion can later be inserted into another sequence using the menu selection *Insert-segment-into-sequence*. The input sequence $x(n)$ does not change after the operation.

Cut Segment from Sequence. This function is used for moving a user-specified range of samples of the input sequence into a new sequence with the name $CLPBRD$. The samples copied are actually removed from the input sequence. This creates a gap in the input sequence. Samples that are on the right side of the gap are pushed to the left to close the gap. For a given input sequence $x(n)$ and a copy range of (N_1, N_2), the output sequence $y(n)$ is computed as

$$y(n) = \begin{cases} x(n), & \text{if } N_1 \leq n \leq N_2 \\ 0, & \text{otherwise} \end{cases}$$

and stored in the data file $CLPBRD$.SEQ. The input sequence is modified as

$$x_{\text{new}}(n) = \begin{cases} x_{\text{old}}(n), & \text{if } n < N_1 \\ x_{\text{old}}(n + N_2 - N_1 + 1), & \text{otherwise} \end{cases}$$

The segment moved from $x(n)$ into $y(n)$ in this fashion can later be inserted into another sequence using the menu selection *Insert-segment-into-sequence*.

Insert Segment into Sequence. This function is used for inserting the contents of the sequence $CLPBRD$ into a specified sequence starting at a specified value of the index. Assume that the clipboard contains the sequence $c(n)$ defined for the range $[N_{c1}, N_{c2}]$. If $c(n)$ is to be inserted into $x(n)$ starting at the index N_1, the latter sequence is modified as

$$x_{\text{new}}(n) = \begin{cases} x_{\text{old}}(n), & n < N_1 \\ c(n - N_1 + N_{c1}), & n = N_1, \ldots, N_{c2} - N_{c1} + N_1 \\ x_{\text{old}}(n - N_{c2} + N_{c1} - 1), & n > N_{c2} - N_{c1} + N_1. \end{cases}$$

A.2.3 Operations Menu

The operations menu contains the basic signal-processing functions. It is further divided into four submenus: *Arithmetic-operations, Nonlinear-operations, Signal-processing-functions,* and *Statistics*.

A.2.3.1 Arithmetic Operations Submenu

The arithmetic operations submenu contains basic arithmetic functions that operate on sequences. These functions include addition, subtraction, multiplication, and division of sequences, complex conjugation, time shifting, real part, imaginary part, even component, and odd component.

Add Two Sequences. Adds two sequences on a sample by sample basis. Given two input sequences $x_1(n)$ and $x_2(n)$, the output sequence $y(n)$ is computed as

$$y(n) = x_1(n) + x_2(n).$$

The range of the output sequence is the union of two input ranges. If the lengths and starting index values of the two input sequences are not equal, then zero padding is performed on the left and/or right of each sequence as necessary.

If both input sequences are real valued, then the resulting output sequence is also real valued. If at least one input sequence is complex, then the result is also complex.

Subtract Sequences. Subtracts a sequence from another on a sample by sample basis. Given two input sequences $x_1(n)$ and $x_2(n)$, the output sequence $y(n)$ is computed as

$$y(n) = x_1(n) - x_2(n).$$

The range of the output sequence is the union of two input ranges. If the lengths and starting index values of the two input sequences are not equal, then zero padding is performed on the left and/or right of each sequence as necessary.

If both input sequences are real valued, then the resulting output sequence is also real valued. If at least one input sequence is complex, then the result is also complex.

Multiply Two Sequences. Multiplies two sequences on a sample by sample basis. Given two input sequences $x_1(n)$ and $x_2(n)$, the output sequence $y(n)$ is computed as

$$y(n) = x_1(n)x_2(n).$$

The range of the output sequence is the union of two input ranges. If the lengths and starting index values of the two input sequences are not equal, then zero padding is performed on the left and/or right of each sequence as necessary.

If both input sequences are real valued, then the resulting output sequence is also real valued. If at least one input sequence is complex, then the result is also complex.

Divide Sequences. Divides each sample of a sequence by the corresponding sample of another sequence. Given two input sequences $x_1(n)$ and $x_2(n)$, the output sequence $y(n)$ is computed as

$$y(n) = \frac{x_1(n)}{x_2(n)}.$$

The range of the output sequence is the union of two input ranges. If the lengths and starting index values of the two input sequences are not equal, then zero padding is performed on the left and/or right of each sequence as necessary.

If a division by zero is encountered, computation is aborted and an error message is displayed. Thus, care should be taken to ensure that the second input sequence does not contain any zero-amplitude samples in the range of the anticipated output sequence.

If both input sequences are real valued, then the resulting output sequence is also real valued. If at least one input sequence is complex, then the result is also complex.

Add Constant Offset. Adds a user-specified complex constant to each sample of the input sequence. Given an input sequence $x(n)$ and a complex constant c, the output sequence is computed as

$$y(n) = x(n) + c.$$

If both $x(n)$ and c are real, then the resulting sequence $y(n)$ is also real. Otherwise, it is complex.

Multiply by Constant. Multiplies each sample of the input sequence with a user-specified complex constant. Given an input sequence $x(n)$, and a complex constant c, the output sequence is computed as

$$y(n) = c\, x(n).$$

If both $x(n)$ and c are real, then the resulting sequence $y(n)$ is also real. Otherwise, it is complex.

Shift Sequence. Shifts the input sequence to the right by a specified number of samples. Left shifts can be achieved by using negative shift amounts. Linear or circular shifts can be obtained by means of option buttons within the dialog window. Given an input sequence $x(n)$ and an integer shift k, the output sequence $y(n)$ is computed as

$$y(n) = x(n - k)$$

if the linear shift option is chosen. Note that, for a finite-length sequence, this simply corresponds to changing the starting index to its new value, and the number of samples does not change. If the range of the input sequence is $[N_1, N_2]$, the output sequence will be in the range $[N_1 + k, N_2 + k]$.

Sec. A.2 Function Reference

If a circular shift is requested, then the output sequence is computed as

$$y(n) = x((n - k))_N,$$

where N is the number of samples in the input sequence and double parentheses indicate modulo operation. This equation also assumes that the input sequence $x(n)$ is defined for $n = 0, \ldots, N - 1$.

It is important to note that, in the case of a circular shift, the input sequence is assumed to start at index 0, even though its actual starting index may be different. In this case, the computed output sequence also starts at index zero. This may be somewhat confusing. As an example, let's assume that the input sequence is specified for the index range $[N_1, N_2]$. First, it is translated to a sequence $\tilde{x}(n)$ as

$$\tilde{x}(n) = x(n + N_1).$$

The resulting $\tilde{x}(n)$ exists for $n = 0, \ldots, N_2 - N_1$. In the second step, the output $y(n)$ is obtained by circularly shifting $\tilde{x}(n)$; that is,

$$y(n) = \tilde{x}((n - k))_N,$$

where $N = N_2 - N_1 + 1$.

Flip Sequence. Flips (time reverses) a sequence. For a given input sequence $x(n)$, the output sequence $y(n)$ is computed as

$$y(n) = x(-n).$$

If the range of the input sequence is $[N_1, N_2]$, the computed output sequence will be in the range $[-N_2, -N_1]$.

Real Part. Computes the real part of the input sequence. For a given input sequence $x(n)$, the output sequence $y(n)$ is

$$y(n) = \mathrm{Re}\{x(n)\}.$$

The output sequence is real and has the same range as the input sequence.

Imaginary Part. Computes the imaginary part of the input sequence. For a given input sequence $x(n)$, the output sequence $y(n)$ is

$$y(n) = \mathrm{Im}\{x(n)\}.$$

The output sequence is real and has the same range as the input sequence. If the input sequence is real, then the output sequence contains zeros.

Complex Conjugate. Computes the complex conjugate of the input sequence. For a given input sequence $x(n)$, the output sequence $y(n)$ is

$$y(n) = x^*(n).$$

Even Component. Computes either the even component or the conjugate-symmetric component of the input sequence. Two check boxes are presented with the titles *Conjugate* and *Modulo-N*. Based on the status of these two check boxes, four distinct formulas are used for computing the output sequence.

1. *Conjugate* = No, *Modulo-N* = No

$$y(n) = \frac{1}{2}[x(n) + x(-n)]$$

2. *Conjugate* = Yes, *Modulo-N* = No

$$y(n) = \frac{1}{2}[x(n) + x^*(-n)]$$

3. *Conjugate* = No, *Modulo-N* = Yes

$$y(n) = \frac{1}{2}[x((n))_N + x((-n))_N]$$

4. *Conjugate* = Yes, *Modulo-N* = Yes

$$y(n) = \frac{1}{2}[x((n))_N + x^*((-n))_N]$$

Note that in the modulo-N computations given in formulas (3) and (4), the starting index of the input sequence is assumed to be zero even if it is not. The computed output sequence also starts at index zero. As an example, assume that the input sequence is specified for the index range $[N_1, N_2]$. First, it is translated to a sequence $\tilde{x}(n)$ as

$$\tilde{x}(n) = x(n + N_1).$$

The resulting $\tilde{x}(n)$ exists for $n = 0, \ldots, N_2 - N_1$. In the second step, the output $y(n)$ is obtained by applying the formulas given to the modified sequence $\tilde{x}(n)$ instead of to the original $x(n)$. The parameter N is computed as

$$N = N_2 - N_1 + 1.$$

Odd Component. Computes either the odd component or the conjugate-antisymmetric component of the input sequence. Two check boxes are presented with the titles *Conjugate* and *Modulo-N*. Based on the status of these two check boxes, four distinct formulas are used for computing the output sequence.

1. *Conjugate* = No, *Modulo-N* = No

$$y(n) = \frac{1}{2}[x(n) - x(-n)]$$

2. *Conjugate* = Yes, *Modulo-N* = No

$$y(n) = \frac{1}{2}[x(n) - x^*(-n)]$$

3. *Conjugate* = No, *Modulo-N* = Yes

$$y(n) = \frac{1}{2}[x((n))_N - x((-n))_N]$$

4. *Conjugate* = Yes, *Modulo-N* = Yes

Sec. A.2 Function Reference

$$y(n) = \frac{1}{2}[x((n))_N - x^*((-n))_N]$$

Note that in the modulo-N computations given in formulas (3) and (4) the starting index of the input sequence is assumed to be zero even if it is not. The computed output sequence also starts at index zero. As an example, assume that the input sequence is specified for the index range $[N_1, N_2]$. First, it is translated to a sequence $\tilde{x}(n)$ as

$$\tilde{x}(n) = x(n + N_1).$$

The resulting $\tilde{x}(n)$ exists for $n = 0, \ldots, N_2 - N_1$. In the second step, the output $y(n)$ is obtained by applying the formulas given to the modified sequence $\tilde{x}(n)$ instead of to the original $x(n)$. The parameter N is computed as

$$N = N_2 - N_1 + 1.$$

A.2.3.2 Nonlinear Operations Submenu

Magnitude. Computes the magnitude of a complex input sequence. For a given input sequence $x(n)$, the output sequence $y(n)$ is computed as

$$y(n) = \{[\text{Re}\{x(n)\}]^2 + [\text{Im}\{x(n)\}]^2\}^{1/2}.$$

The output sequence is real, and its range is the same as that of the input sequence. If the input sequence is real, then its magnitude is equal to its absolute value, that is,

$$y(n) = |x(n)|.$$

Phase. Computes the phase of a complex input sequence. For a given input sequence $x(n)$, the output sequence $y(n)$ is computed as

$$y(n) = \tan^{-1}\left[\frac{\text{Im}\{x(n)\}}{\text{Re}\{x(n)\}}\right].$$

The output sequence is real, and its range is the same as that of the input sequence. Phase angles computed are in radians. If the input sequence is real, then the output sequence samples take on the values 0 or π depending on the signs of the input sequence samples.

Cartesian Form. This function is used for converting a sequence from polar to Cartesian form. For a given input sequence $x(n)$, the output sequence $y(n)$ is computed as

$$y(n) = \text{Re}\{x(n)\} \cos[\text{Im}\{x(n)\}] + j\,\text{Re}\{x(n)\} \sin[\text{Im}\{x(n)\}].$$

The range of the output sequence is the same as the range of the input sequence. If the input sequence is real, the computed output sequence will be equal to the input sequence.

Polar Form. This function is used for converting a sequence from Cartesian to polar form. For a given input sequence $x(n)$, the output sequence $y(n)$ is computed as

$$y(n) = |x(n)| + j\,\text{arg}[x(n)].$$

The range of the output sequence is the same as the range of the input sequence. If the input sequence is real and nonnegative, the computed output sequence will be equal to the input sequence.

Square. Computes the square of the input sequence. For a given input sequence $x(n)$, the output sequence $y(n)$ is computed as

$$y(n) = [x(n)]^2.$$

The range of the output sequence is the same as the range of the input sequence. If the input sequence is real, the computed output sequence will also be real and nonnegative. If the input sequence is complex, then the computed sequence will be complex. Using this function on a sequence $x(n)$ is equivalent to using the *Multiply sequences* function of the *Arithmetic operations* submenu and specifying $x(n)$ for both inputs.

Square Root. Computes the square root of the input sequence. For a given input sequence $x(n)$, the output sequence $y(n)$ is computed as

$$y(n) = \sqrt{x(n)}.$$

The range of the output sequence is the same as the range of the input sequence. If the input sequence is real and nonnegative, the computed output sequence will also be real and nonnegative. If the input sequence is complex or if it contains negative amplitude samples, then the computed sequence will be complex. Note that the square root of a complex number $re^{j\theta}$ is computed as

$$\sqrt{re^{j\theta}} = \sqrt{r}e^{j\theta/2}$$

and the square-root of a sequence is computed by applying this to each sample of the sequence.

Reciprocate. Computes the reciprocal of the input sequence. For a given input sequence $x(n)$, the output sequence $y(n)$ is computed as

$$y(n) = \frac{1}{x(n)}.$$

The range of the output sequence is the same as the range of the input sequence. The input sequence $x(n)$ must not have any zero-valued samples; otherwise, a floating-point overflow will occur. If this happens, try adding a very small positive number to each sample in the input sequence before computing its reciprocal; that is, compute

$$y(n) \approx \frac{1}{x(n) + \epsilon}$$

where $\epsilon \ll |x(n)|$ for all nonzero samples of $x(n)$.

Raise to Power. Raises a sequence to a specified power. For a given input sequence $x(n)$, the output sequence $y(n)$ is computed as

$$y(n) = [x(n)]^c,$$

where c is a real-valued constant. The range of the output sequence is the same as the range of the input sequence. If the constant c is negative, then the input sequence $x(n)$ must not have any zero-valued samples; otherwise, a floating-point overflow will occur. If this happens, try adding a very small positive number to each sample in the input sequence before computing $y(n)$; that is, compute

Sec. A.2 Function Reference

$$y(n) \approx [x(n) + \epsilon]^c,$$

where $\epsilon \ll |x(n)|$ for all nonzero samples of $x(n)$.

Exponential. Computes the exponential function of a sequence. For a given input sequence $x(n)$, the output sequence $y(n)$ is computed as

$$y(n) = e^{x(n)}.$$

The range of the output sequence is the same as the range of the input sequence. If the input sequence is real, the computed output sequence will also be real; otherwise, it will be complex.

Logarithm. Computes the logarithm function of a sequence. Either the natural logarithm or the base 10 logarithm can be computed. For a given input sequence $x(n)$, the output sequence $y(n)$ is computed as

$$y(n) = \ln[x(n)]$$

or

$$y(n) = \log_{10}[x(n)],$$

depending on the user's choice. The range of the output sequence is the same as the range of the input sequence. If the input sequence is real and nonnegative, the computed output sequence will be real; otherwise, it will be complex. Note that the natural logarithm of a complex number is computed as

$$\ln(re^{j\theta}) = \ln(r) + j\theta.$$

If the base 10 logarithm is desired, the only difference will be a constant scale factor. Care should be taken to ensure that $r \neq 0$ for all samples of the sequence; otherwise, a floating-point error will result.

Compander. This function is used for applying μ-law or A-law companding formulas to the samples of a specified real-valued discrete sequence. It can be used in conjunction with the *Quantize-sequence* menu selection to simulate the behavior of pulse-code modulation (PCM) systems. For a given input sequence $x(n)$, the output sequence $y(n)$ is computed using one of the following formulas:

1. Compressor, μ-law:

$$y = \frac{\text{sgn}(x)}{\ln(1 + \mu)} \ln\left(1 + \mu \frac{|x|}{x_{\max}}\right)$$

2. Expander, μ-law:

$$y = \frac{x_{\max} \text{sgn}(x)}{\mu}[(1 + \mu)^{|x|} - 1]$$

3. Compressor, A-law:

$$y = \begin{cases} \dfrac{A}{1 + \ln(A)}\left(\dfrac{x}{x_{\max}}\right), & 0 \leq \dfrac{|x|}{x_{\max}} \leq \dfrac{1}{A} \\ \dfrac{\text{sgn}(x)}{1 + \ln(A)}\left[1 + \ln\left(A\dfrac{|x|}{x_{\max}}\right)\right], & \dfrac{1}{A} \leq \dfrac{|x|}{x_{\max}} \leq 1 \end{cases}$$

4. Expander, A-law:

$$y = \begin{cases} \dfrac{x_{max} x[1 + \ln(A)]}{A}, & 0 \leq |x| \leq \dfrac{1}{1 + \ln(A)} \\ \dfrac{x_{max} \text{sgn}(x)\exp\{|x| + |x|\ln(A) - 1\}}{A}, & \dfrac{1}{1 + \ln(A)} \leq |x| \leq 1 \end{cases}$$

Note that the above formulas require all samples of $x(n)$ to be in the range $[-x_{max}, x_{max}]$. Any samples of $x(n)$ that are outside this range are set equal to $\pm x_{max}$. The sign is determined based on which limit is closer. The input sequence $x(n)$ must be real. If it is complex, an error message will be generated.

Quantize Sequence. Computes a quantized version of a real-valued input sequence. The user specifies the number of quantization levels M and the amplitude limits x_{min} and x_{max}. Each sample of the input signal is quantized according to the following:

$$\text{Quantization step size: } \Delta = \frac{x_{max} - x_{min}}{M}$$

$$\text{Quantization levels: } q_i = x_{min} + \frac{\Delta}{2} + i\Delta, \quad i = 0, \ldots, M - 1$$

$$y(n) = q_i \text{ such that } |x(n) - q_i| \leq |x(n) - q_j|, \quad j = 0, \ldots, M - 1$$

The input sequence $x(n)$ must be real. If it is complex, an error message will be generated.

Limit Sequence. This function is used for limiting amplitude values to a specified range. For a given input sequence $x(n)$ and amplitude limits x_{min} and x_{max}, the output sequence $y(n)$ is computed as

$$y(n) = \begin{cases} x(n), & x_{min} \leq x(n) \leq x_{max} \\ x_{min}, & x(n) < x_{min} \\ x_{max}, & x(n) > x_{max} \end{cases}$$

The input sequence $x(n)$ must be real. If it is complex, an error message will be generated.

Trigonometric Functions. Computes a specified trigonometric function of the input sequence. For a given input sequence $x(n)$, the output sequence $y(n)$ is computed as

$$y(n) = \text{fnc}[x(n)],$$

where the function fnc[] is one of the following: cos[], sin[], tan[], arccos[], arcsin[], and arctan[]. The range of the computed output sequence is the same as the range of the input sequence. The arguments to the functions cos[], sin[], and tan[] are assumed to be in radians. Return values of the functions arccos[], arcsin[], and arctan[] are also in radians. The input sequence used may be complex. The following formulas are used in computing trigonometric functions of complex numbers:

$$\cos(\alpha + j\beta) = \cos(\alpha)\cosh(\beta) - j\sin(\alpha)\sinh(\beta)$$

$$\sin(\alpha + j\beta) = \sin(\alpha)\cosh(\beta) + j\cos(\alpha)\sinh(\beta)$$

$$\tan(\alpha + j\beta) = \frac{\sin(\alpha + j\beta)}{\cos(\alpha + j\beta)}$$

$$\arccos(x) = -j \ln\left(x + \sqrt{x^2 - 1}\right)$$

$$\arcsin(x) = -j \ln\left(jx + \sqrt{1 - x^2}\right)$$

$$\arctan(x) = -j\frac{1}{2}\ln\left(\frac{1 + jx}{1 - jx}\right)$$

Hyperbolic Functions. Computes a specified hyperbolic function of the input sequence. For a given input sequence $x(n)$, the output sequence $y(n)$ is computed as

$$y(n) = \text{fnc}[x(n)],$$

where the function fnc[] is one of the following: cosh[], sinh[], tanh[], arccosh[], arcsinh[], and arctanh[]. The range of the computed output sequence is the same as the range of the input sequence. The input sequence used may be complex. The following formulas are used in computing hyperbolic functions of complex numbers:

$$\cosh(x) = \frac{1}{2}[e^x + e^{-x}]$$

$$\sinh(x) = \frac{1}{2}[e^x - e^{-x}]$$

$$\tanh(x) = \frac{\sinh(x)}{\cosh(x)}$$

$$\text{arccosh}(x) = \ln\left(x + \sqrt{x^2 - 1}\right)$$

$$\text{arcsinh}(x) = \ln\left(x + \sqrt{x^2 + 1}\right)$$

$$\text{arctanh}(x) = \frac{1}{2}\ln\left(\frac{1 + x}{1 - x}\right)$$

A.2.3.3 Processing Functions Submenu

Difference Equation. This menu selection is used for evaluating a difference equation recursively. Both linear and nonlinear difference equations are supported. The output sequence is computed within a specified range of the index n by recursing through the difference equation. For example, consider the difference equation

$$y(n) = 0.7y(n - 1) + 0.2y(n - 2) + x(n) - x(n - 2).$$

To evaluate $y(n)$, enter the right side of the difference equation into the equation field as

```
0.7*y(n - 1) + 0.2*y(n - 2) + x(n) - x(n - 2)
```

A nonlinear difference equation such as

$$y(n) = 0.5y(n - 1) + 0.5\frac{x(n)}{y(n - 1)}$$

can be specified as

$$0.5*y(n - 1) + 0.5*x(n)/y(n - 1)$$

Note that only the right side of the difference equation is entered, and the left side is assumed to be y(n). The index variable must be n. The equation entered must be syntactically correct in terms of parentheses matching and function arguments. The precedence of arithmetic operators is the same as in most programming languages (for example, C, Pascal, or FORTRAN), and parentheses can be used to change the order of operators. The following arithmetic operators are supported:

+	(addition)
−	(subtraction)
*	(multiplication)
/	(division)
^	(raise to power)

Complex numbers can be entered by using braces. Inside braces, real and imaginary parts of a number are separated by a comma. For example, the expression

$$\{1,2\}$$

represents the complex number $1 + j2$. Real and imaginary parts of a complex number can be mathematical expressions themselves. For example,

$$\{cos(0.2*pi*n), sin(0.2*pi*n)\}$$

is valid. The intrinsic functions supported are as follows:

sin(x)	cos(x)	tan(x)	atan(x)	atan2(x, y)
sinh(x)	cosh(x)	tanh(x)	exp(x)	ln(x)
log(x)	abs(x)	real(x)	imag(x)	round(x)
trunc(x)	sqr(x)	sqrt(x)	imp(n)	step(n)
sign(x)	gau(m, v)	uni(l, u)		

For detailed definitions of these functions, see the discussion of the formula-entry method in Section A.2.2.1.

Convolution. Computes the convolution of two sequences. Given two input sequences $x_1(n)$ and $x_2(n)$, the output sequence $y(n)$ is computed as

$$y(n) = \sum_k x_1(k)x_2(n - k).$$

If the nonzero ranges of the input sequences are $[N_1, N_2]$ and $[N_3, N_4]$, respectively, the range of the computed output sequence is $[N_1 + N_3, N_2 + N_4]$. If both input sequences are real valued, then the resulting output sequence is also real valued. If at least one input sequence is complex, then the result is also complex. Internally, PC-DSP utilizes the fast Fourier transform (FFT) for fast computation of the convolution sum.

Sec. A.2 Function Reference

Autocorrelation. Computes the autocorrelation function of a sequence. If the input sequence is assumed to be a signal from a stationary random process, the computed output sequence corresponds to an estimate of the autocorrelation function of that process. Biased or unbiased estimation formulas can be used. For a given input sequence $x(n)$ defined in the range $[0, N - 1]$, the output sequence is computed as

$$y(n) = \frac{1}{N} \sum_{k=0}^{N-1-|n|} x^*(k) x(k+n) \quad \text{(biased formula)}$$

or

$$y(n) = \frac{1}{N - |n|} \sum_{k=0}^{N-1-|n|} x^*(k) x(k+n) \quad \text{(unbiased formula)}.$$

The range of the output sequence $y(n)$ is $[-N + 1, N - 1]$. If the starting index of $x(n)$ is not zero, then the limits of the summation are adjusted to cover all nonzero values of the product $x^*(k)x(k + n)$. If both input sequences are real valued, then the resulting output sequence is also real valued. If at least one of the input sequences is complex, then the result is also complex. Internally, PC-DSP utilizes the fast Fourier transform (FFT) for fast computation of the autocorrelation function.

Autocovariance. Computes the autocovariance function of a sequence. If the input sequence is assumed to be a signal from a stationary random process, the computed output sequence corresponds to an estimate of the autocovariance function of that process. Biased or unbiased estimation formulas can be used. For a given input sequence $x(n)$ defined in the range $[0, N - 1]$, the output sequence is computed as

$$y(n) = \frac{1}{N} \sum_{k=0}^{N-1-|n|} [x(k) - \hat{\mu}_x]^* [x(k+n) - \hat{\mu}_x] \quad \text{(biased formula)}$$

or

$$y(n) = \frac{1}{N - |n|} \sum_{k=0}^{N-1-|n|} [x(k) - \hat{\mu}_x]^* [x(k+n) - \hat{\mu}_x] \quad \text{(unbiased formula)},$$

where $\hat{\mu}_x$ is an estimate of the mean and is computed as

$$\hat{\mu}_x = \frac{1}{N} \sum_{n=0}^{N-1} x(n).$$

The range of the output sequence $y(n)$ is $[-N + 1, N - 1]$. If the starting index of $x(n)$ is not zero, then the limits of the preceding summations are adjusted to cover all nonzero values of the relevant terms. If both input sequences are real valued, then the resulting output sequence is also real valued. If at least one input sequence is complex, then the result is also complex. Internally, PC-DSP utilizes the fast Fourier transform (FFT) for fast computation of the autocovariance function.

Cross Correlation. Computes the cross-correlation function of two sequences. If the two input sequences are assumed to be signals from stationary random processes, the computed output sequence corresponds to an estimate of the cross-correlation function of those processes. Biased or unbiased estimation formulas can be used. Given two input sequences $x_1(n)$ and $x_2(n)$ defined for index ranges $[0, N_1 - 1]$ and $[0, N_2 - 1]$, respectively, the output sequence is computed as

$$y(n) = \frac{1}{N_1} \sum_{k=0}^{N_1 - 1 - |n|} x_1^*(k) x_2(k + n) \quad \text{(biased formula)}$$

or

$$y(n) = \frac{1}{N_1 - |n|} \sum_{k=0}^{N_1 - 1 - |n|} x_1^*(k) x_2(k + n) \quad \text{(unbiased formula)}$$

The range of the output sequence is $[-N_1 + 1, N_2 - 1]$. If the starting index values of $x_1(n)$ and $x_2(n)$ are not zero, then the limits of the summation and the starting index value of the output $y(n)$ are adjusted to cover all nonzero values of the product $x^*_1(k)x_2(k + n)$. If both input sequences are real valued, then the resulting output sequence is also real valued. If at least one input sequence is complex, then the result is also complex. Internally, PC-DSP utilizes the fast Fourier transform (FFT) for fast computation of the cross-correlation function.

Cross Covariance. Computes the cross-covariance function of two sequences. If the two input sequences are assumed to be signals from stationary random processes, the computed output sequence corresponds to an estimate of the cross-covariance function of those processes. Biased or unbiased estimation formulas can be used. Given two input sequences $x_1(n)$ and $x_2(n)$ defined for index ranges $[0, N_1 - 1]$ and $[0, N_2 - 1]$, respectively, the output sequence is computed as

$$y(n) = \frac{1}{N_1} \sum_{k=0}^{N_1 - 1 - |n|} [x_1(k) - \hat{\mu}_{x1}]^* [x_2(k + n) - \hat{\mu}_{x2}] \quad \text{(biased formula)}$$

or

$$y(n) = \frac{1}{N_1 - |n|} \sum_{k=0}^{N_1 - 1 - |n|} [x_1(k) - \hat{\mu}_{x1}]^* [x_2(k + n) - \hat{\mu}_{x2}] \quad \text{(unbiased formula)},$$

where $\hat{\mu}_{x1}$ and $\hat{\mu}_{x2}$ are estimates of the mean and are computed as

$$\hat{\mu}_{x1} = \frac{1}{N_1} \sum_{n=0}^{N_1 - 1} x_1(n)$$

and

$$\hat{\mu}_{x2} = \frac{1}{N_2} \sum_{n=0}^{N_2 - 1} x_2(n).$$

The range of the output sequence is $[-N_1 + 1, N_2 - 1]$. If the starting index values of $x_1(n)$ and $x_2(n)$ are not zero, then the limits of the preceding summations and the starting index value of the output $y(n)$ are adjusted to cover all nonzero values of the relevant terms. If both input sequences are real valued, then the resulting output sequence is also real valued. If at least one of the input sequences is complex, then the result is also complex. Internally, PC-DSP utilizes the fast Fourier transform (FFT) for fast computation of the cross-covariance function.

Sec. A.2 Function Reference

Downsample Sequence. Computes a downsampled version of the input sequence. For a given input sequence $x(n)$ and a downsampling rate M, the output sequence $y(n)$ is computed as

$$y(n) = x(nM).$$

Upsample Sequence. Computes an upsampled version of the input sequence. For a given input sequence $x(n)$ and an upsampling rate L, the output sequence $y(n)$ is computed as

$$y(n) = \begin{cases} x(n/L), & \text{if } n = kL,\ k\text{: integer} \\ 0, & \text{otherwise.} \end{cases}$$

A.2.3.4 Statistics Submenu

Signal Statistics. Computes and displays various signal statistics for a specified discrete-time signal. The statistics computed are as follows:

Sample mean:
$$\tilde{m}_x = \frac{1}{N} \sum_n x(n)$$

Sample variance:
$$\hat{\sigma}_x^2 = \frac{1}{N-1} \left[\sum_n x^2(n) - \frac{1}{N} \left(\sum_n x(n) \right)^2 \right]$$

Median: d_x such that $\text{Prob}(x(n) \leq d_x) \approx 0.5$

Sum: $s_1 = \sum_n x(n)$

Absolute sum: $s_2 = \sum_n |x(n)|$

Mean-square value: $\text{MSV} = \frac{1}{N} \sum_n |x(n)|^2$

Histogram. Computes the histogram of a specified sequence. The user specifies lower and upper limits $x_{\min}$ and $x_{\max}$ as well as the number of histogram bins M. For a given input sequence $x(n)$, an M-sample output sequence $y(n)$ is generated such that

$$\Delta = \frac{x_{\max} - x_{\min}}{M},$$

$y(i) = $ number of samples in $x(n)$ that satisfy

$$x_{\min} + i\Delta < x(n) \leq x_{\min} + (i+1)\Delta.$$

Before the histogram is computed, samples of the input sequence are clipped to fit into the specified amplitude range $(x_{\min}, x_{\max})$. As a result of this, samples with amplitudes that are outside the specified amplitude range contribute to the leftmost or the rightmost bins of the histogram.

A.2.4 Transforms Menu

FFT. Computes the fast Fourier transform (FFT) of a sequence. For a given input sequence $x(n)$ and a selected window function $w(n)$, the output sequence $y(k)$ is computed as

$$y(k) = \sum_{n=0}^{N-1} x(n)w(n)e^{-j(2\pi kn/N)}, \qquad k = 0, \ldots, N-1.$$

It is important to note that the input sequence is assumed to start at index 0, even though its actual starting index may be different. As an example, let's assume that the input sequence is specified for the index range $[N_1, N_2]$. First, it is translated to a sequence $\tilde{x}(n)$ as

$$\tilde{x}(n) = x(n + N_1).$$

The resulting sequence $\tilde{x}(n)$ exists for $n = 0, \ldots, N_2 - N_1$. The previous FFT equation is applied to this modified sequence.

The window function $w(n)$ has the same number of samples as $x(n)$. It can be one of the following: rectangular, triangular, Hamming, Hanning, Blackman, Kaiser, or Chebyshev. The choice of the rectangular window corresponds to a straight FFT with no window function. For theoretical details of window functions, refer to the section on Fourier series design of FIR filters.

The FFT size N is specified by the user. It must be an integer power of 2 since PC-DSP uses the radix-2 decimation-in-time algorithm for computing the FFT. If N is greater than the number of samples in the sequence $x(n)$, then the sequence is padded with zeros to extend its length to N. If N is less than the number of samples in $x(n)$, an error message is generated.

Inverse FFT. Computes the inverse fast Fourier transform of a complex sequence. For a given input sequence $x(k)$, the output sequence $y(n)$ is computed as

$$y(n) = \frac{1}{N} \sum_{k=0}^{N-1} x(k)e^{j(2\pi nk/N)}, \qquad n = 0, \ldots, N-1.$$

It is important to note that the input sequence is assumed to start at index 0, even though its actual starting index may be different. As an example, let's assume that the input sequence is specified for the index range $[N_1, N_2]$. First, it is translated to a sequence $\tilde{x}(k)$ as

$$\tilde{x}(k) = x(k + N_1).$$

The resulting sequence $\tilde{x}(k)$ exists for $k = 0, \ldots, N_2 - N_1$. The previous inverse FFT equation is applied to this modified sequence.

The inverse-FFT size N is specified by the user. It must be an integer power of 2 since PC-DSP uses the radix-2 decimation-in-time algorithm for computing the inverse FFT. If N is greater than the number of samples in the sequence $x(k)$, then the sequence is padded with zeros to extend its length to N. If N is less than the number of samples in $x(k)$, an error message is generated.

DFT. Computes the discrete Fourier transform (DFT) of a sequence. For a given input sequence $x(n)$ and a selected window function $w(n)$, the output sequence $y(k)$ is computed as

$$y(k) = \sum_{n=0}^{N-1} x(n)w(n)e^{-j(2\pi kn/N)}, \qquad k = 0, \ldots, N-1.$$

Sec. A.2 Function Reference

In contrast with the FFT function, this function computes the transform by direct application of this equation. Therefore, it is significantly slower than the FFT function. On the other hand, the DFT size N does not have to be a power of 2; any integer that is equal to greater than the number of samples in the input signal can be used.

It is important to note that the input sequence is assumed to start at index 0, even though its actual starting index may be different. As an example, let's assume that the input sequence is specified for the index range $[N_1, N_2]$. First, it is translated to a sequence $\tilde{x}(n)$ as

$$\tilde{x}(n) = x(n + N_1).$$

The resulting sequence $\tilde{x}(n)$ exists for $n = 0, \ldots, N_2 - N_1$. The previous DFT equation given is applied to this modified sequence.

The window function $w(n)$ has the same number of samples as $x(n)$. It can be one of the following: rectangular, triangular, Hamming, Hanning, Blackman, Kaiser, or Chebyshev. The choice of the rectangular window corresponds to a straight DFT with no window function. For theoretical details of window functions, refer to the section on Fourier series design of FIR filters.

The DFT size N is specified by the user. If N is greater than the number of samples in the sequence $x(n)$, then the sequence is padded with zeros to extend its length to N. If N is less than the number of samples in $x(n)$, an error message is generated.

Inverse DFT. Computes the inverse discrete Fourier transform of a complex sequence. For a given input sequence $x(k)$, the output sequence $y(n)$ is computed as

$$y(n) = \frac{1}{N} \sum_{k=0}^{N-1} x(k) e^{j(2\pi nk/N)}, \qquad n = 0, \ldots, N-1$$

In contrast with the inverse-FFT function, this function computes the inverse transform by direct application of this equation. Therefore, it is significantly slower than the inverse-FFT function. On the other hand, the inverse-DFT size N does not have to be a power of 2; any integer that is equal to greater than the number of samples in the input signal can be used.

It is important to note that the input sequence is assumed to start at index 0, even though its actual starting index may be different. As an example, let's assume that the input sequence is specified for the index range $[N_1, N_2]$. First, it is translated to a sequence $\tilde{x}(k)$ as

$$\tilde{x}(k) = x(k + N_1).$$

The resulting sequence $\tilde{x}(k)$ exists for $k = 0, \ldots, N_2 - N_1$. The previous inverse-DFT equation is applied to this modified sequence.

The inverse-DFT size N is specified by the user. If N is greater than the number of samples in the sequence $x(k)$, then the sequence is padded with zeros to extend its length to N. If N is less than the number of samples in $x(k)$, an error message is generated.

DTFT. Computes the discrete-time Fourier transform of a complex sequence. For a given input sequence $x(n)$, the DTFT is computed as

$$X(\omega) = \sum_{n=-N_1}^{N_2} x(n) e^{-j\omega n}.$$

The resulting transform is saved as a continuous signal rather than a sequence. (The filename extension of the disk file created is TRN, as opposed to SEQ). PC-DSP evaluates the DTFT at N equally spaced frequencies in the range $-\pi \leq \omega < \pi$ using fast Fourier transform techniques. The number of transform samples is chosen as follows:

1. If the length of the sequence $x(n)$ is less than or equal to 512, then $N = 512$.
2. If the length of the sequence $x(n)$ is greater than 512, then N is set equal to the smallest integer power of 2 that is greater than or equal to the number of samples in $x(n)$.

Once the DTFT is computed, it can be graphed with the *F2* key or tabulated with the *F3* key. Alternatively, *Tabulate-transform* and *Plot-transform* menu choices (explained later) can be used.

List Transforms. Provides a list of discrete-time Fourier transforms stored in the current data directory. (This directory is specified using the *Directories* option of the *Configuration Options* submenu. See Section A.2.1.1.) The *current transform* is also shown. (The current transform is usually the result of the most recent operation. Recall that this is the transform that is graphed with the *F2* key and tabulated with the *F3* key.) Any transform in the list can be made current by double clicking on its name. Without a mouse, the direction keys and the ‹ENTER› key can be used instead.

Tabulate Transform. Allows on-screen tabulation of a specified discrete-time Fourier transform. A new window is created with the values of the transform. The transform is tabulated in the range of the angular frequency $-\pi \leq \omega < \pi$. More than one transform tabulation can be on the screen at one time. Tabulation windows can be sized, moved, or closed using the corresponding menu selections in the *Editor* submenu. (See Section A.2.1.2.)

Plot Transform. Generates an on-screen plot of a specified discrete-time Fourier transform in the range of the angular frequency $-\pi \leq \omega < \pi$. If the transform is complex, its magnitude and phase components will be shown separately. While the plot is on the screen, some keys have special meanings:

    C: Continuous plot
    D: Discrete plot
    G: Toggle grid on and off
    M: Toggle between Cartesian and polar representations
    L: Toggle between linear and logarithmic (dB) scale for the magnitude plot

Delete Transform. This menu choice is used for deleting a specified discrete-time Fourier transform from the disk. The user is asked to confirm this decision before the transform is deleted.

Inverse z-Transform. Computes the inverse z-transform of a rational function $X(z)$ using power series (long division) method. The user specifies the transform by means of numerator and denominator coefficients. Additionally, the desired range of the sample index is specified. Consider a rational function $X(z)$ in the form

$$X(z) = \frac{b_q z^q + \cdots + b_1 z + b_0}{a_p z^p + \cdots + a_1 z + a_0},$$

where the leading denominator coefficient is nonzero; that is, $a_p \neq 0$. The coefficients of the transform are entered in descending order:

Numerator: $\quad b_q, \ldots, b_1, b_0$
Denominator: $\quad a_p, \ldots, a_1, a_0$

Left-sided or right-sided solutions can be obtained. For the right-sided solution, long division is carried out using the coefficients as given. If the left-sided solution is requested, numerator and denominator polynomials are ordered with ascending powers of the complex variable z before the long division is performed.

A.2.5 Filters Menu

List Filters. Provides a list of filters (analog and discrete) stored in the current data directory. (This directory is specified using the *Directories* option of the *Configuration Options* submenu. See Section A.2.1.1.) Recall that PC-DSP stores filter data in binary files with the file-name extension FLT.

Delete Filter. This menu choice is used for deleting a specified filter data file from the disk. The user is asked to confirm this decision before the filter is deleted.

Analog Filter Design. This function facilitates the design of frequency-selective analog filters. The specifications of the desired analog filter are entered using the conventions outlined in Appendix B. Following is a summary of design parameters:

Filter type: Low pass, high pass, bandpass, band reject.

Approximation: Butterworth, Chebyshev type 1, Chebyshev type 2 (inverse Chebyshev), or elliptic. For details, see Appendix B.

Specifications: Basically, in analog filter design, there are three parameters that are dependent on each other. These are the passband tolerance (Δ_1 or R_p), the stopband tolerance (Δ_2 or A_s), and the filter order N. Only two of these three parameters can be specified by the user; the third parameter is determined by the program as a function of the first two. Thus, the tolerance scheme can be specified in three different forms:

1. Specify the passband and the stopband tolerances, and the filter order will be chosen to be the minimum that satisfies the requirements. This can be used for all approximation types.
2. Specify the filter order and the passband tolerance. This choice is available for all approximation types except Chebyshev type 2 filters.
3. Specify the filter order and the stopband tolerance. This choice is available for Butterworth and Chebyshev type 2 filters.

Critical frequencies: Critical frequencies are entered in hertz. The number of critical frequencies needed depends on the filter type, as well as on which tolerance scheme is used.

Excess tolerance use: If the desired filter tolerance scheme is specified in terms of passband and stopband tolerance values (specification type 1), PC-DSP designs the lowest-order filter that satisfies the requirements. The theoretical minimum filter order computed is typically a noninteger value that is rounded to the next integer. This rounding results in a filter that exceeds the specifications. The user can control how this excess is used. Three choices are possible:
1. Use excess tolerance in the passband.
2. Use excess tolerance in the stopband.
3. Distribute the excess tolerance equally (on a decibel scale) between the passband and the stopband.

Frequency Transformations. This menu selection is used for transforming an analog low-pass filter into an analog high-pass, bandpass, or band-reject filter by means of a frequency transformation applied to the low-pass transfer function. The user selects the type of transformation desired, and the geometric center frequency, in hertz, that is to be used in the transformation. For theoretical details on frequency transformations, see Appendix B.

IIR Filter Design. This function facilitates the design of frequency-selective IIR discrete-time filters from analog prototypes using the bilinear transformation method. Other design methods (impulsive invariance, first backward difference, and matched z-transform) are available using the step by step design approach explained in Chapter 8.

The specifications of the desired IIR filter are entered using the conventions outlined in Chapter 8. Following is a summary of design parameters:

Filter type: Low pass, high pass, bandpass, and band reject.

Approximation: Butterworth, Chebyshev type I, Chebyshev type II (inverse Chebyshev), or elliptic. For details, see Appendix B.

Specifications: The tolerance scheme can be specified in three different forms:
1. Specify the passband and the stopband tolerances, and the filter order will be chosen to be the minimum that satisfies the requirements. This can be used for all approximation types.
2. Specify the filter order and the passband tolerance. This choice is available for all approximation types except Chebyshev type II filters.
3. Specify the filter order and the stopband tolerance. This choice is available for Butterworth and Chebyshev type II filters.

Critical frequencies: Critical frequencies F_i are entered as dimensionless normalized quantities. These are obtained through division of actual frequencies by the sampling frequency so that the normalized value of the sampling frequency is 1. The number of critical frequencies needed depends on the filter type, as well as on which tolerance scheme is used.

Excess tolerance use: If the desired filter tolerance scheme is specified in terms of passband and stopband tolerance values (specification type 1), PC-DSP designs the lowest-order filter that satisfies the requirements. The theoretical minimum filter order computed is typically a noninteger value that is rounded to the next integer. This rounding results in a filter that exceeds the specifications. The user can control how this excess is used. Three choices are possible:
1. Use excess tolerance in the passband.
2. Use excess tolerance in the stopband.
3. Distribute the excess tolerance equally (on a decibel scale) between the passband and the stopband.

Transform Analog Filter. This function produces a discrete-time filter from an analog prototype filter using one of the following methods:

1. Bilinear transformation
2. Impulse invariance
3. First backward difference
4. Matched z-transform

Theoretical details on these four methods can be found in Chapter 8. The user specifies the desired method and the sampling interval (in seconds). The analog prototype filter may have been designed using the analog filter design function of PC-DSP, or it may have been entered into PC-DSP using the *Enter-external-filter* function.

Analyze Filter. This menu choice is used for analyzing filters that are either designed using PC-DSP or entered into PC-DSP using the *Enter-external-filter* function. The following analysis types are available:

1. Magnitude
2. Decibel magnitude
3. Phase
4. Time delay
5. Group delay
6. Impulse response
7. Poles and zeros
8. Filter coefficients

After one of the analysis types is selected, a tabular listing of its values or a screen graph can be obtained by pressing the appropriate button. Note that certain analysis types are only available for certain types of filters. For example, analysis type 8 is not available for FIR filters, and analysis type 6 is not available for analog filters.

It is possible to save a selected analysis function, for example, decibel magnitude response, into a PC-DSP data file in the appropriate format by pressing the *Save* button. The *Report* button displays the design report that is generated for filters designed using PC-DSP.

Simulate Filter. Simulates the processing of a specified discrete-time sequence through a specified IIR or FIR filter. The names of the input sequence and the discrete-time filter are specified by the user. For IIR filters, a cascade implementation consisting of direct-form-II blocks is utilized with all initial filter states set equal to zero. For FIR filters, FFT-based fast convolution technique is used. The number of output samples computed is equal to the number of input samples.

Enter External Filter. Allows the description of an analog or discrete-time filter to be entered interactively and to be converted to a filter data file in PC-DSP format. This menu choice is primarily used for analyzing and simulating a pencil and paper design using the analysis and simulation functions of PC-DSP.

Analog filters and discrete-time IIR filters can be described either in terms of their transfer function coefficients or of their z-domain poles and zeros. If the transfer function approach is used, numerator and denominator coefficients are entered in descending powers of z and are separated from each other with semicolons. If poles and zeros are used, complex poles and zeros must occur in conjugate pairs.

FIR filters are described using either z-domain zero locations or impulse response samples. Complex zeros must occur in conjugate pairs.

A.2.5.1 FIR Filter Design Submenu

Fourier Series Method. This menu choice is used for designing FIR filters using the Fourier series design method with window functions. Fourier series design method is based on computing the impulse response of the ideal filter to be approximated and then windowing it to obtain a finite-length sequence as the impulse response of an FIR filter. For theoretical details, refer to Chapter 8.

The types of filters that can be designed using the Fourier series method are low-pass, high-pass, bandpass, and band-reject filters, differentiators, and Hilbert transform filters. For frequency-selective filters, critical frequencies of the desired filter are entered using normalized values. All edge frequencies are normalized with respect to the sampling frequency. Hence, the largest edge frequency that can be used is 0.5. Available window functions are rectangular, Bartlett (triangular), Hanning, Hamming, and Blackman windows.

Kaiser Window Design. This is a variant of the Fourier-series design method that uses the Kaiser window. Recall that the Kaiser window has a variable parameter β to adjust the trade-off between the main-lobe width and the side-lobe attenuation of the window spectrum. In FIR filter design based on the Kaiser window, the parameter β and the filter-length N are determined using empirical formulas due to Kaiser [16]:

$$\beta = \begin{cases} 0.1102(A_s - 8.7), & A_s > 50 \\ 0.5842(A_s - 21)^{0.4} + 0.07886(A_s - 21), & 21 \leq A_s \leq 50 \\ 0, & A_s < 21 \end{cases}$$

and

$$N - 1 = \frac{A_s - 8}{2.285 \Delta \omega}$$

where A_s is the desired stopband attenuation in decibels and $\Delta\omega$ is the width of the transition band.

Low-pass, high-pass, bandpass, and band-reject filters can be designed. Critical frequencies of the desired filter are entered using normalized values. All edge frequencies are normalized with respect to the sampling frequency. Hence, the largest edge frequency that can be used is 0.5.

Frequency-sampling Method. Designs frequency-sampling FIR filters with raised-cosine-shaped transition regions. The frequency-sampling technique for FIR filter design is based on sampling the desired frequency response at N equally spaced values of the angular frequency, computing the inverse DFT of the resulting sequence, and circularly shifting the inverse-DFT result. For theoretical details, refer to Chapter 8.

Multiband filters, differentiators, and Hilbert transform filters can be designed using this method. The desired frequency-domain behavior is specified in terms of the normalized edge frequencies of each band and the desired magnitude in each band. All edge frequencies are normalized with respect to the sampling frequency. Hence, the largest edge frequency that can be used is 0.5. For multiband filters, the lower edge of the first frequency band and the upper edge of the last frequency band must be 0.0 and 0.5, respectively. This requirement does not apply to differentiators and Hilbert transform filters.

Frequency-sampling filters with no transition bands can also be designed with this method. To achieve this, specify edge frequencies such that no frequency samples fall within the transition band.

Parks–McClellan Method. Facilitates FIR filter design using the design algorithm that is due to Parks and McClellan. The filters designed using this algorithm are optimum in the sense that the maximum ripple (the difference between actual and desired responses) is minimized over a set of disjoint frequency bands. Three types of filters can be designed using the Parks–McClellan algorithm:

1. Multiband filters
2. Differentiators
3. Hilbert transform filters

Up to five frequency bands can be used for a multiband filter. Differentiators and Hilbert transform filters use only one frequency band.

The desired frequency-domain behavior is specified in terms of the normalized edge frequencies of each band, the desired magnitude in each band, and the relative weight factor of each band. Note that frequency bands must be disjoint; that is, there must be transition (don't care) bands between them. All edge frequencies are normalized with respect to the sampling frequency. Hence, the largest edge frequency that can be used is 0.5. For multiband filters, the lower edge of the first frequency band and the upper edge of the last frequency band must be 0.0 and 0.5, respectively. This requirement does not apply to differentiators and Hilbert transform filters.

The weight factor for each band must be positive. For a differentiator, the desired value in a frequency band is interpreted as the desired slope in that band.

A.2.6 Spectrum Menu

Periodogram: Computes the power spectral density (PSD) estimate of a stochastic process based on a finite-length data record from that process using Welch's method of averaging modified periodograms. The input sequence is broken into segments that may or may not overlap. A selected window function is applied to each segment, and a periodogram is computed for each windowed segment. These periodograms are scaled and averaged to obtain an estimate of the power spectral density of the process.

The user specifies the length of each segment and the amount of overlap (if any) between consecutive segments. N samples of the PSD estimate are computed in the normalized frequency range $-0.5 \leq F < 0.5$. The number of PSD samples, N, is the FFT size used for each segment. It is specified by the user and must not be less than the length of each segment of the data record.

Both real and complex sequences can be used as input. The PSD estimate can be computed on either a linear or logarithmic (dB) scale. If it is computed in decibels, it can be optionally scaled so that the peak value is normalized to 0 dB. Note that the standard periodogram (without Welch's modifications) can be obtained by specifying the segment length to be equal to the length of the input sequence and selecting the rectangular window.

Blackman–Tukey Method. Computes the Blackman–Tukey PSD estimate for a given finite-length data record. In this method, autocorrelation lag estimates $\hat{r}_{xx}(m)$ are computed from the input sequence for the range $-M \leq m \leq M$, using either the biased estimator or the unbiased estimator. Afterward, the DTFT of the estimated autocorrelation sequence is computed as an estimate of the power spectral density. If desired, a window function may be applied to the estimated autocorrelation sequence before computing the DTFT.

The user specifies the largest autocorrelation lag to be used (M) and the choice of the window function to be applied to the autocorrelation sequence. Then 512 samples of the PSD estimate are computed for the normalized frequency range $-0.5 \leq F < 0.5$.

Both real and complex sequences can be used as input. The PSD estimate can be computed on either a linear or logarithmic (dB) scale. If it is computed in decibels, it can be optionally scaled so that the peak value is normalized to 0 dB.

Note that, in general, the Blackman–Tukey PSD estimate may exhibit negative values for some frequency ranges. Only the use of a Bartlett (triangular) window guarantees a nonnegative PSD estimate.

Autoregressive. In autoregressive (AR) modeling of data, the process is assumed to have been generated by driving an all-pole linear system with white noise. An all-pole system with M poles is characterized by a transfer function in the form

$$H(z) = \frac{1}{1 + \sum_{k=1}^{M} a_k z^{-k}}.$$

If the input to this system is a white noise process $w(n)$ with zero mean and variance σ^2, the power spectral density of the resulting output process $x(n)$ is

Sec. A.2 Function Reference

$$P_x(F) = \frac{1}{\left|1 + \sum_{k=1}^{M} a_k e^{-j2\pi Fk}\right|^2} \sigma^2.$$

A PSD estimate is obtained by estimating the coefficients $a_1, a_2, \ldots, a_M$ and the driving noise variance σ^2 from the available data record. PC-DSP implements two methods for estimation of AR coefficients:

1. Yule–Walker method
2. Burg method

Using either method, any of the following can be computed:

1. PSD estimate
2. AR coefficient estimates for a specified model order M
3. Reflection coefficient estimates up to a specified model order M
4. Driving noise variance estimates up to a specified model order M

The maximum AR model order is 200. If a PSD estimate is requested, 512 samples are computed for the normalized frequency range $-0.5 \leq F < 0.5$.

Both real and complex sequences can be used as input. The PSD estimate can be computed on either a linear or logarithmic (dB) scale. If it is computed in decibels, it can be optionally scaled so that the peak value is normalized to 0 dB.

Moving Average. In contrast to AR modeling, a moving average (MA) model is an all-zero model for which the process is assumed to have been generated by driving a FIR filter with white noise; that is,

$$x(n) = w(n) + \sum_{k=1}^{M} b_k w(n-k)$$

with the PSD given by

$$P_x(F) = \left|1 + \sum_{k=1}^{M} b_k e^{-j2\pi Fk}\right|^2 \sigma^2.$$

A PSD estimate is obtained by estimating the coefficients $b_1, b_2, \ldots, b_M$ and the driving noise variance σ^2 from the available N-sample data record. PC-DSP uses Durbin's method to estimate the coefficients of the MA model. This method consists of two steps. In the first step, the Yule–Walker method is used to model the input sequence as an Lth order AR process, where $M \ll L \ll N$. In the second step, the coefficients of this Lth order AR model are used as input data to compute a set of M coefficients $b_1, b_2, \ldots, b_M$, again using the Yule–Walker method.

Either the PSD estimate or the MA model coefficients may be requested. If the PSD estimate is requested, 512 samples are computed for the normalized frequency range $-0.5 \leq F < 0.5$.

Both real and complex sequences can be used as input. The PSD estimate can be computed on either a linear or logarithmic (dB) scale. If it is computed in decibels, it can be optionally scaled so that the peak value is normalized to 0 dB.

ARMA. An autoregressive moving average (ARMA) process is generated by driving a pole–zero filter with white noise. The z-domain system function of the filter is in the form

$$H(z) = \frac{1 + \sum_{k=1}^{Q} b_k z^{-k}}{1 + \sum_{k=1}^{P} a_k z^{-k}}.$$

The PSD of the output process is

$$P_x(F) = \left| \frac{1 + \sum_{k=1}^{Q} b_k e^{-j2\pi Fk}}{1 + \sum_{k=1}^{P} a_k e^{-j2\pi Fk}} \right| \sigma^2.$$

Finding an optimum ARMA model estimate for a given process is computationally quite difficult since the equations to be solved are nonlinear. PC-DSP uses an ad hoc method for finding a "nearly optimum" solution. First, the AR coefficients $a_1, a_2, \ldots, a_P$ are estimated using the Yule–Walker method. The data record $x(n)$ is then filtered through a FIR filter with the system function

$$A(z) = 1 + \sum_{k=1}^{P} a_k z^{-k}.$$

Finally, the MA coefficients $b_1, b_2, \ldots, b_Q$ are estimated from the output sequence of the filter, using Durbin's method.

Either the PSD estimate or the MA model coefficients may be requested. If the PSD estimate is requested, 512 samples are computed for the normalized frequency range $-0.5 \leq F < 0.5$.

Both real and complex sequences can be used as input. The PSD estimate can be computed on either a linear or logarithmic (dB) scale. If it is computed in decibels, it can be optionally scaled so that the peak value is normalized to 0 dB.

A.3 DATA FILE FORMATS

Sequence File Structure

Entry	Data Type	Bytes
Signature	String (zero terminated)	
Comment	String (zero terminated)	Variable
Sequence name	String (zero terminated)	Variable
Type[a]	Integer	2
Length	Integer	2
Starting index (N_1)	Integer	2
Real part of $x(N_1)$	Real	8
Imaginary part of $x(N_1)$	Real[b]	8
Real part of $x(N_1 + 1)$	Real	8
Imaginary part of $x(N_1 + 1)$	Real[b]	8
.	.	.
.	.	.
.	.	.

[a] Sequence type: 1 = real sequence; 2 = complex sequence.
[b] For complex sequence only.

Signal File Structure

Entry	Data Type	Bytes
Signature	String (zero terminated)	
Comment	String (zero terminated)	Variable
X-axis title	String (zero terminated)	Variable
Sequence name	String (zero terminated)	Variable
Type[a]	Integer	2
Length	Integer	2
Starting index (N_1)	Integer	2
X increment (Δx)	Real	8
Real part of $x(N_1\Delta x)$	Real	8
Imaginary part of $x(N_1\Delta x)$	Real[b]	8
Real part of $x[(N_1 + 1)\Delta x]$	Real	8
Imaginary part of $x[(N_1 + 1)\Delta x]$	Real[b]	8
.	.	.
.	.	.
.	.	.

[a] Signal type: 1 = real signal; 2 = complex signal.
[b] For complex signal only.

B

Analog Filter Design

In Chapter 7, analog filter prototypes were used for the design of IIR digital filters. The design procedure was based on first finding the s-domain transfer function $G(s)$ of an appropriate analog filter and then converting it into the desired digital filter transfer function $H(z)$ by means of a transformation between the s-plane and the z-plane. At that point, we assumed that the analog prototypes needed were readily available. This appendix will outline the procedures for designing analog filters using Butterworth, Chebyshev, and elliptic approximations.

Specifications of analog filters to be designed are usually given in terms of a set of tolerance limits, similar to the IIR digital filter specifications we considered in Chapter 7. Specification diagrams for the four frequency-selective filter types are shown in Fig. B.1.

In analog filter design problems, first a low-pass filter is designed regardless of the type of filter that is required. Once an appropriate low-pass filter is obtained, it can be easily converted to any of the other filter types (high pass, bandpass, or band reject) by employing a *frequency transformation*. Because of this, we will mostly concentrate on the design of analog low-pass filters. Frequency transformations will be considered in Section B.5.

Tolerance limits for the desired analog filter magnitude response can also be given on a logarithmic (dB) scale as shown in Fig. B.2. The maximum allowed decibel passband ripple R_p and the minimum required decibel stopband attenuation A_s are related to Δ_1 and Δ_2 by

$$R_p = 20 \log\left(\frac{1}{1 - \Delta_1}\right) \tag{B.1}$$

and

$$A_s = 20 \log\left(\frac{1}{\Delta_2}\right). \tag{B.2}$$

Sec. B.1 Butterworth Filters

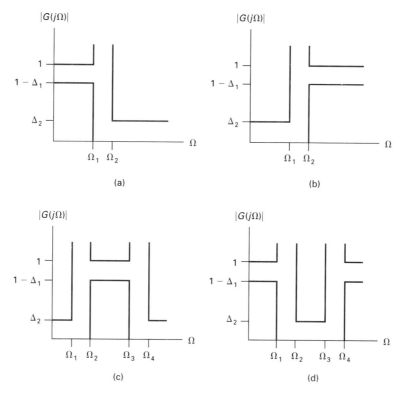

Figure B.1 Specification diagrams for frequency-selective filters: (a) low pass; (b) high pass; (c) bandpass; (d) band reject.

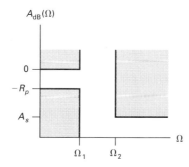

Figure B.2 Decibel tolerance specifications for an analog low-pass filter.

B.1 BUTTERWORTH FILTERS

A Butterworth analog low-pass filter is characterized by the squared-magnitude function

$$|G(j\Omega)|^2 = \frac{1}{1 + (\Omega/\Omega_c)^{2N}}, \tag{B.3}$$

where the two parameters Ω_c and N completely describe the filter. The magnitude response of the Butterworth filter can be graphed from (B.3) by evaluating the square root. (See

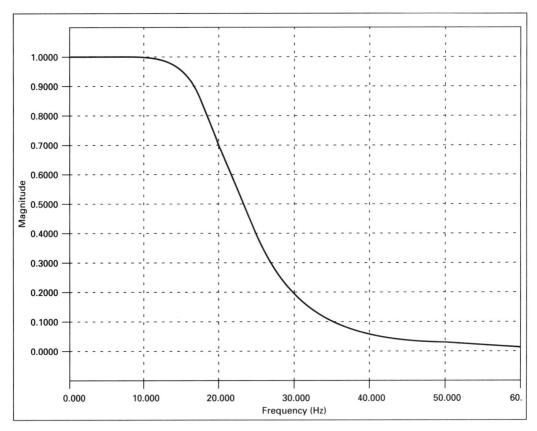

Figure B.3 Butterworth filter magnitude characteristic.

Fig. B.3.) It can be easily verified that the magnitude is equal to $1/\sqrt{2}$ at the point where $\Omega = \Omega_c$. Since this approximately corresponds to the -3-dB point on a logarithmic scale, the parameter Ω_c is referred to as the *3-dB cutoff frequency* of the filter. The integer parameter N is the *order* of the filter.

Given values of the two filter parameters Ω_c and N, the design problem requires finding the s-domain transfer function $G(s)$ the squared magnitude of which matches the right side of (B.3). For a transfer function $G(s)$ with real coefficients, it can be shown that

$$|G(j\Omega)|^2 = G(s)G(-s)|_{s = j\Omega}. \qquad (B.4)$$

To obtain $G(s)$, the procedure in (B.4) needs to be reversed. Substituting s for $j\Omega$ or, equivalently, $-s^2$ for Ω^2 in (B.3), we obtain

$$G(s)G(-s) = \frac{1}{1 + (-s^2/\Omega_c^2)^N}. \qquad (B.5)$$

If $G(s)$ has a pole in the left half s-plane at $s = \alpha + j\beta$, then $G(-s)$ has a corresponding pole at $s = -\alpha - j\beta$; thus, the poles of $G(s)$ are mirror images of the poles of $G(-s)$

Sec. B.1 Butterworth Filters

with respect to the origin. To extract $G(s)$ from the product $G(s)G(-s)$, we need to find the poles of this product and separate the two sets of poles that belong to $G(s)$ and $G(-s)$, respectively. The cases of even and odd filter orders N need to be treated separately.

If N Is Odd: The poles of the product $G(s)G(-s)$ are the values of s that satisfy the equation

$$s^{2N} - \Omega_c^{2N} = 0,$$

which can be written in the alternative form

$$s^{2N} = \Omega_c^{2N} e^{j2\pi k}$$

and can be solved for s, yielding

$$s_k = \Omega_c e^{j(k\pi/N)}, \quad \text{for } k = 0, \ldots, 2N - 1. \tag{B.6}$$

A few observations are in order: All the poles of the product $G(s)G(-s)$ are located on a circle with radius equal to Ω_c. Furthermore, the poles are equally spaced on the circle, and the angle between any two adjacent poles is $2\pi/N$. Since $G(s)$ and $G(-s)$ have only real coefficients, all complex poles appear in complex conjugate pairs. Figure B.4 depicts the situation for $\Omega_c = 2$ rad/s and $N = 5$. We will associate the poles on the left half s-plane with $G(s)$ and the other poles with $G(-s)$. For the resulting filter to be stable, it is necessary to use the poles on the left side for $G(s)$. The transfer function $G(s)$ for the Butterworth filter can now be constructed in the form

$$G(s) = \frac{\Omega_c^N}{\Pi_k (s - s_k)}, \quad \text{for all } k \text{ that satisfy } \frac{\pi}{2} < \frac{k\pi}{N} < \frac{3\pi}{2}. \tag{B.7}$$

If N Is Even.   The poles of the product $G(s)G(-s)$ are the solutions of the equation

$$s^{2N} + \Omega_c^{2N} = 0,$$

which can be written as

$$s^{2N} = \Omega_c^{2N} e^{j\pi(2k + 1)}.$$

In this case, the general solution for the poles is

$$s_k = \Omega_c e^{j[\pi(2k + 1)/2N]}, \quad \text{for } k = 0, \ldots, 2N - 1. \tag{B.8}$$

For even filter orders, poles of the magnitude-squared transfer function are still equally spaced on a circle. Compared to odd filter orders, the only difference in this case is that no poles appear on the real axis. The transfer function $G(s)$ can be constructed in the same way as was done for odd values of N.

In most cases, the desired behavior of a low-pass filter is specified in terms of the critical frequencies Ω_1 and Ω_2 and the decibel tolerance values R_p and A_s. It is necessary to determine the corresponding values of Ω_c and N so that the filter can be designed. The conditions at critical frequencies can be expressed as inequalities:

$$-10 \log\left[1 + \left(\frac{\Omega_1}{\Omega_c}\right)^{2N}\right] \geq -R_p \tag{B.9}$$

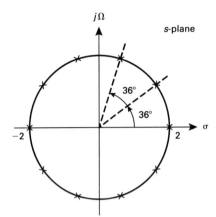

Figure B.4 Poles of $G(s)G(-s)$ for $\Omega_c = 2$ rad/s and $N = 5$.

and

$$-10 \log\left[1 + \left(\frac{\Omega_2}{\Omega_c}\right)^{2N}\right] \leq -A_s. \quad \text{(B.10)}$$

Equations (B.9) and (B.10) can be rearranged to yield

$$\left(\frac{\Omega_1}{\Omega_c}\right)^{2N} \leq 10^{R_p/10} - 1 \quad \text{(B.11)}$$

and

$$\left(\frac{\Omega_2}{\Omega_c}\right)^{2N} \geq 10^{A_s/10} - 1. \quad \text{(B.12)}$$

Dividing both sides of (B.11) by the corresponding terms of (B.12), we obtain

$$\left(\frac{\Omega_1}{\Omega_2}\right)^{2N} \leq \frac{10^{R_p/10} - 1}{10^{A_s/10} - 1},$$

which can be solved for N, resulting in

$$N \geq \frac{\log\sqrt{(10^{R_p/10} - 1)/(10^{A_s/10} - 1)}}{\log(\Omega_1/\Omega_2)}. \quad \text{(B.13)}$$

The smallest value of N that satisfies the inequality in (B.13) is typically a noninteger and must be rounded up to the next integer. This rounding of N results in the tolerance specifications being exceeded; that is, the designed filter will do better than the specifications require. To exceed the specifications in the stopband, we can solve the passband equation (B.11) for Ω_c, obtaining

$$\Omega_c = \frac{\Omega_1}{(10^{R_p/10} - 1)^{1/2N}}. \quad \text{(B.14)}$$

Sec. B.1 Butterworth Filters

This choice of Ω_c would result in (B.11) being satisfied with an equality, while (B.12) remains as an inequality. Alternatively, the specifications can be exceeded in the passband by solving the stopband equation (B.12), which yields

$$\Omega_c = \frac{\Omega_2}{(10^{A_s/10} - 1)^{1/2N}}. \tag{B.15}$$

It is also possible to distribute the excess tolerance equally between the two critical frequencies by rewriting (B.9) and (B.10) as equalities with the use of an intermediate variable Λ as

$$-10 \log\left[1 + \left(\frac{\Omega_1}{\Omega_c}\right)^{2N}\right] \geq -R_p + \Lambda \tag{B.16}$$

and

$$-10 \log\left[1 + \left(\frac{\Omega_2}{\Omega_c}\right)^{2N}\right] \leq -A_s - \Lambda \tag{B.17}$$

and solving both equations simultaneously for Ω_c and Λ.

In some design problems, the desired filter order might be specified in advance. In this case, only one critical frequency and its corresponding tolerance value can be specified. As an example, if N, Ω_1, and R_p are given, the following inequality holds:

$$-10 \log[1 + \left(\frac{\Omega_1}{\Omega_c}\right)^{2N}] \geq -R_p. \tag{B.18}$$

If the 3-dB cutoff frequency Ω_c can be found, the filter transfer function can be constructed. In the worst case, (B.18) can be taken as an equality to obtain

$$\left(\frac{\Omega_1}{\Omega_c}\right)^{2N} = 10^{R_p/10} - 1 = K_1, \tag{B.19}$$

which can be solved for Ω_c to yield

$$\Omega_c = \frac{\Omega_1}{10^{\log(K_1)/2N}} \tag{B.20}$$

If the stopband attenuation A_s is specified instead of the passband ripple R_p, then the inequality to be satisfied is

$$-10 \log\left[1 + \left(\frac{\Omega_2}{\Omega_c}\right)^{2N}\right] \leq -A_s. \tag{B.21}$$

Taking (B.21) with equality, we obtain

$$\left(\frac{\Omega_2}{\Omega_c}\right)^{2N} = 10^{A_s/10} - 1 = K_2.$$

Solving for Ω_c,

$$\Omega_c = \frac{\Omega_2}{10^{\log(K_2)/2N}}. \qquad (B.22)$$

B.2 CHEBYSHEV FILTERS

The squared-magnitude function for a Chebyshev type I lowpass filter is

$$|G(j\Omega)|^2 = \frac{\alpha}{1 + \epsilon^2 C_N^2(\Omega/\Omega_c)} \qquad (B.23)$$

where α and ϵ are positive constants and $C_N(x)$ represents the Chebyshev polynomial of order N. Given the parameter values α, ϵ, N, and Ω_c, the design procedure can be broken into the following steps:

1. Find the Nth-order Chebyshev polynomial $C_N(x)$.
2. Replace the independent variable x by Ω/Ω_c to obtain $C_N(\Omega/\Omega_c)$.
3. Form the squared-magnitude function as given by (B.23).
4. Find the poles of $G(s)$ following the same procedure that was used in the previous section for Butterworth filter design. [Substitute $s^2 = -\Omega^2$, find the poles on the s-plane, select the ones on the left half s-plane, and construct $G(s)$.] The main difference is that the poles of the squared-magnitude function will no longer be on a circle.

Before we proceed with the steps outlined, we will briefly review the properties of Chebyshev polynomials.

Chebyshev Polynomials

The denominator of the squared-magnitude function for a Chebyshev filter requires the Nth-order Chebyshev polynomial, which is defined as

$$C_N(x) = \cos(N \cos^{-1}(x)). \qquad (B.24)$$

The definition in (B.24) is somewhat easier to visualize if it is broken up into two parts:

$$x = \cos(\theta)$$

and
$$C_N(x) = \cos(N\theta). \qquad (B.25)$$

Thus, for a specified value of N, $C_N(x)$ can be obtained by first expressing $\cos(N\theta)$ in terms of $\cos(\theta)$ (using trigonometric identities) and then replacing each occurrence of $\cos(\theta)$ with x. It can easily be shown that

$$C_0(x) = 1$$

and
$$C_1(x) = x.$$

Sec. B.2 Chebyshev Filters

To find the second-order Chebyshev polynomial, we need to express $\cos(2\theta)$ in terms of $\cos(\theta)$. The trigonometric identity needed here is

$$\cos(2\theta) = 2\cos^2(\theta) - 1$$

and the second-order Chebyshev polynomial is

$$C_2(x) = 2x^2 - 1.$$

Higher-order Chebyshev polynomials can also be obtained in this fashion; however, as the order increases, the procedure would become increasingly tedious. An easier method is to use the recursion formula

$$C_{N+1}(x) = 2xC_N(x) - C_{N-1}(x), \tag{B.26}$$

which allows any order of Chebyshev polynomial to be computed if the polynomials of the previous two orders are known. Knowing $C_0(x)$ and $C_1(x)$, the next polynomial $C_2(x)$ can be computed as

$$\begin{aligned} C_2(x) &= 2xC_1(x) - C_0(x) \\ &= 2x^2 - 1, \end{aligned}$$

and $C_3(x)$ can be computed as

$$\begin{aligned} C_3(x) &= 2xC_2(x) - C_1(x) \\ &= 4x^3 - 3x. \end{aligned}$$

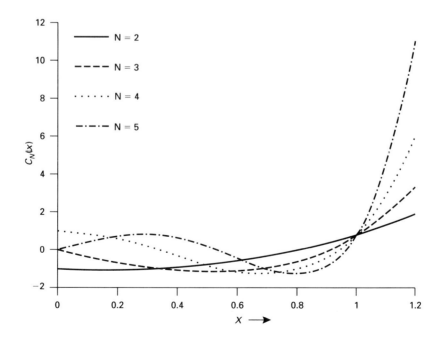

Figure B.5 Some Chebyshev polynomials.

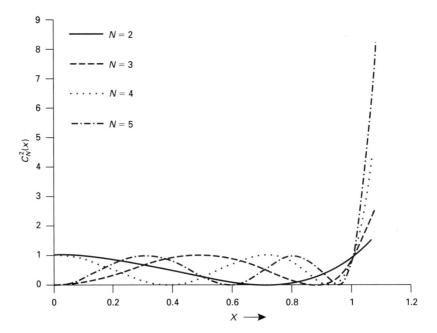

Figure B.6 Squared Chebyshev polynomials.

Figure B.5 shows the graphs of the first few Chebyshev polynomials. The following common properties can be readily observed:

1. All Chebyshev polynomials are equal to 1 for $x = 1$; that is, $C_N(1) = 1$ for all N.
2. If $|x| < 1$, then $|C_N(x)| < 1$ for all Chebyshev polynomials.
3. For $|x| > 1$, all Chebyshev polynomials grow without bound.
4. At $x = 0$, odd-order Chebyshev polynomials are equal to zero, and even-order Chebyshev polynomials are equal to ± 1.

Figure B.6 shows the squared Chebyshev polynomials for the first few orders. It is interesting to note that, with the exception of $C_0(x)$, all squared Chebyshev polynomials look like low-pass filter characteristics turned upside down. For $-1 \leq x \leq 1$, all squared Chebyshev polynomials oscillate in the range [0, 1]. They grow monotonically for values of x outside this range. It is this behavior that makes their use in filter design possible. The magnitude response of a Chebyshev low-pass filter is shown in Fig. B.7.

Poles of the Chebyshev Filter

To construct the transfer function $G(s)$ for a Chebyshev filter, the poles of the product $G(s)G(-s)$ are needed. Recall that the product $G(s)G(-s)$ is obtained by replacing $j\Omega$ with s in the squared-magnitude function $|G(j\Omega)|^2$ of Eq. (B.23). Thus, the poles of $G(s)G(-s)$ are the solutions of the equation

Sec. B.2 Chebyshev Filters

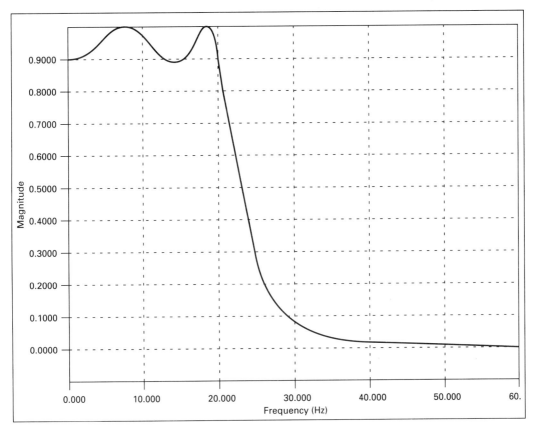

Figure B.7 Chebyshev filter magnitude response.

$$1 + \epsilon^2 C_N^2(s_k/j\Omega_c) = 0 \tag{B.27}$$

for $k = 0, \ldots, 2N - 1$. Let

$$x_k = \frac{s_k}{j\Omega_c}.$$

Then

$$1 + \epsilon^2 C_N^2(x_k) = 0. \tag{B.28}$$

Using the definition of the Chebyshev polynomial, this can be rewritten as

$$1 + \epsilon^2 \cos^2(N\theta_k) = 0, \tag{B.29}$$

where $x_k = \cos(\theta_k)$. Solving (B.29) for $\cos(N\theta_k)$, we obtain

$$\cos(N\theta_k) = \pm j\frac{1}{\epsilon}. \tag{B.30}$$

Let $\theta_k = \alpha_k + j\beta_k$, where α_k and β_k are both real. Using the appropriate trigonometric identity, (B.30) can be written as

$$\cos(N\alpha_k)\cos(jN\beta_k) - \sin(N\alpha_k)\sin(jN\beta_k) = \pm j\frac{1}{\epsilon}.$$

Recognizing that $\cos(jN\beta_k) = \cosh(N\beta_k)$ and $\sin(jN\beta_k) = j\sinh(N\beta_k)$,

$$\cos(N\alpha_k)\cosh(N\beta_k) - j\sin(N\alpha_k)\sinh(N\beta_k) = \pm j\frac{1}{\epsilon}. \quad (B.31)$$

Equating real and imaginary parts of both sides in (B.31) yields

$$\cos(N\alpha_k)\cosh(N\beta_k) = 0 \quad (B.32)$$

and

$$\sin(N\alpha_k)\sinh(N\beta_k) = \pm\frac{1}{\epsilon}. \quad (B.33)$$

From (B.32), we obtain

$$\cos(N\alpha_k) = 0$$

and

$$\alpha_k = \frac{(2k+1)\pi}{2N}. \quad (B.34)$$

Using this value of α_k in (B.33) results in

$$\sin(N\alpha_k) = \pm 1,$$

and

$$\beta_k = \frac{\sinh^{-1}(1/\epsilon)}{N}. \quad (B.35)$$

Once α_k and β_k are found, the poles s_k can be computed from

$$\begin{aligned} s_k &= j\Omega_c x_k \\ &= j\Omega_c \cos(\theta_k) \\ &= j\Omega_c \{\cos(\alpha_k)\cosh(\beta_k) - j\sin(\alpha_k)\sinh(\beta_k)\}. \end{aligned} \quad (B.36)$$

Obtaining N and ϵ for Chebyshev Filters

In some design problems, it may be desirable to obtain the lowest-order filter that satisfies certain tolerance limits. We will assume that the desired behavior is specified in terms of the critical frequencies Ω_c and Ω_2 and the decibel tolerance values R_p and A_s. We will derive the equations for determining ϵ and N.

At the passband edge frequency $\Omega = \Omega_c$, we have

$$20\log|G(j\Omega_c)| = -R_p.$$

Sec. B.2 Chebyshev Filters

Recall that $C_N(1) = 1$ for all Chebyshev polynomials. Thus, we can write

$$10 \log\left(\frac{1}{1 + \epsilon^2}\right) = -R_p,$$

which can be solved for ϵ to yield

$$\epsilon = \sqrt{10^{R_p/10} - 1}. \tag{B.37}$$

Now we need to determine the filter order N. At the stopband edge frequency Ω_2, we have the following inequality:

$$20 \log|G(j\Omega_2)| \le -A_s. \tag{B.38}$$

To simplify the notation, we will define $\Omega_n = \Omega_2/\Omega_c$. Now (B.38) can be expressed as

$$10 \log\left(\frac{1}{1 + \epsilon^2 C_N^2(\Omega_n)}\right) \le -A_s. \tag{B.39}$$

In the worst case, (B.39) can be taken with equality and solved for $C_N(\Omega_n)$ to yield

$$C_N(\Omega_n) = \sqrt{\frac{10^{A_s/10} - 1}{10^{R_p/10} - 1}}, \tag{B.40}$$

where we have also used the value of ϵ given by (B.37). Let

$$F = \sqrt{\frac{10^{A_s/10} - 1}{10^{R_p/10} - 1}} \tag{B.41}$$

and

$$\cos(\theta_n) = \Omega_n. \tag{B.42}$$

Then

$$\cos(N\theta_n) = F. \tag{B.43}$$

Since θ_n is complex, we can express it in Cartesian form as $\theta_n = \alpha_n + j\beta_n$, and (B.43) becomes

$$\cos(N\alpha_n)\cosh(N\beta_n) - j\sin(N\alpha_n)\sinh(N\beta_n) = F. \tag{B.44}$$

Since the right side of (B.44) is real, α_n must be equal to zero, and we have

$$\cosh(N\beta_n) = F. \tag{B.45}$$

Using (B.42),

$$\beta_n = \frac{\theta_n}{j} = \cosh^{-1}(\Omega_n).$$

Equation (B.45) can now be solved for the filter order N.

$$N = \frac{\cosh^{-1}(F)}{\cosh^{-1}(\Omega_n)}. \tag{B.46}$$

The filter-order obtained through (B.46) may not necessarily be an integer and must be rounded up to the next integer. Any excess tolerance that results from rounding up the filter order will be used toward improving the stopband response. If it is desired to use the excess tolerance to improve the passband tolerance instead, then the parameter ϵ must be recomputed using the filter order just found in conjunction with (B.39); thus,

$$C_N^2(\Omega_n) = \frac{10^{A_s/10} - 1}{\epsilon^2}$$

and
$$\epsilon = \frac{\sqrt{10^{A_s/10} - 1}}{C_N(\Omega_n)}. \tag{B.47}$$

B.3 INVERSE CHEBYSHEV FILTERS

The squared-magnitude function for an inverse Chebyshev low-pass filter (also referred to as a type II Chebyshev filter) is

$$|G(j\Omega)|^2 = \frac{\epsilon^2 C_N^2(\Omega_s/\Omega)}{1 + \epsilon^2 C_N^2(\Omega_s/\Omega)}, \tag{B.48}$$

where ϵ is a positive constant, and $C_N(x)$ represents the Chebyshev polynomial of order N. The parameter Ω_s is the stopband edge frequency. Figure B.8 shows the typical magnitude response of a low-pass inverse Chebyshev filter. In contrast to the type I Chebyshev filter response shown in Fig. B.7, the magnitude characteristic of the inverse (type II) Chebyshev filter is smooth in the passband and exhibits equiripple behavior in the stopband.

Poles of the Inverse Chebyshev Filter

The denominator of the squared-magnitude function $|G(j\Omega)|^2$ in (B.48) is very similar to that of the type I Chebyshev filter given by (B.23). As a result, we will be able to use most of the results of the previous section in determining the poles of the inverse Chebyshev filter. Recall that the product $G(s)G(-s)$ is obtained by replacing $j\Omega$ with s in the squared-magnitude function $|G(j\Omega)|^2$. Thus, for an inverse low-pass Chebyshev filter, the poles of $G(s)G(-s)$ are the solutions of the equation

$$1 + \epsilon^2 C_N^2\left(\frac{j\Omega_s}{s_k}\right) = 0 \tag{B.49}$$

for $k = 0, \ldots, 2N - 1$. Let

$$x_k = \frac{j\Omega_s}{s_k}.$$

Then

$$1 + \epsilon^2 C_N^2(x_k) = 0. \tag{B.50}$$

Sec. B.3 Inverse Chebyshev Filters

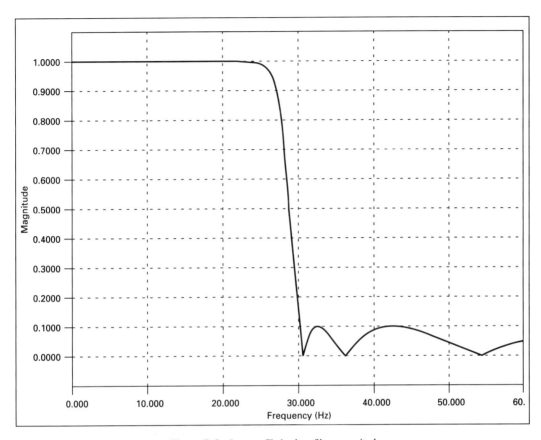

Figure B.8 Inverse Chebyshev filter magnitude response.

Note that this is identical to (B.28) used in computing the poles of the type I Chebyshev filter; therefore, the solution that was obtained for x_k can be used here as well:

$$x_k = \cos(\theta_k)$$

with

$$\alpha_k = \text{Re}\{\theta_k\} = \frac{(2k+1)\pi}{2N},$$

and

$$\beta_k = \text{Im}\{\theta_k\} = \frac{\sinh^{-1}(1/\epsilon)}{N}.$$

Once α_k and β_k are found, the poles s_k can be computed from

$$s_k = \frac{j\Omega_s}{x_k} = \frac{j\Omega_s}{\cos(\alpha_k + j\beta_k)} \quad (B.51)$$

$$= j\Omega_s \{\cos(\alpha_k)\cosh(\beta_k) - j\sin(\alpha_k)\sinh(\beta_k)\}^{-1}.$$

Obtaining N and ϵ for Inverse Chebyshev Filters

The parameters Ω_s, ϵ, and N are all that is needed to uniquely describe an inverse Chebyshev low-pass filter. If the desired filter is specified in terms of critical frequencies Ω_1 and Ω_s and the decibel tolerance values R_p and A_s, we need to obtain ϵ and N in order to proceed with the design.

Evaluating the squared-magnitude square function (B.48) at the stopband edge frequency $\Omega = \Omega_s$, and remembering that $C_N(1) = 1$ for all Chebyshev polynomials, we obtain

$$10 \log\left(\frac{\epsilon^2}{1 + \epsilon^2}\right) = -A_s,$$

which can be solved for ϵ to yield

$$\epsilon = \frac{1}{\sqrt{10^{A_s/10} - 1}}. \tag{B.52}$$

Now we need to determine the filter order N. At the passband edge frequency Ω_1, we have the following inequality:

$$20 \log|G(j\Omega_1)| \geq -R_p. \tag{B.53}$$

To simplify the notation, we will define $\Omega_n = \Omega_s/\Omega_1$. Now we can write

$$10 \log\left[\frac{\epsilon^2 C_N^2(\Omega_n)}{1 + \epsilon^2 C_N^2(\Omega_n)}\right] \geq -R_p. \tag{B.54}$$

In the worst case, (B.54) can be taken with equality and solved for $C_N(\Omega_n)$ to yield

$$C_N(\Omega_n) = \sqrt{\frac{10^{A_s/10} - 1}{10^{R_p/10} - 1}}, \tag{B.55}$$

where we have also used the value of ϵ given by (B.52). It is interesting to note that this last equation is identical to Eq. (B.40) obtained in the previous section for type I Chebyshev filters. The solution for N must also be in the same form as that of the previous section and is repeated here for convenience:

$$N = \frac{\cosh^{-1}\left(\sqrt{(10^{A_s/10} - 1/(10^{R_p/10} - 1)}\right)}{\cosh^{-1}(\Omega_n)}. \tag{B.56}$$

The filter order obtained through (B.56) may not necessarily be an integer and must be rounded up to the next integer. Any excess tolerance that results from rounding up will be used toward an improved passband response. If it is desired to use the excess tolerance in the stopband instead, then the parameter ϵ must be recomputed using the filter order just found in conjunction with (B.54); that is,

$$C_N^2(\Omega_n) = \frac{10^{A_s/10} - 1}{\epsilon^2}$$

and

$$\epsilon = \frac{\sqrt{10^{A_s/10} - 1}}{C_N(\Omega_n)}. \tag{B.57}$$

Sec. B.4   Elliptic Filters   205

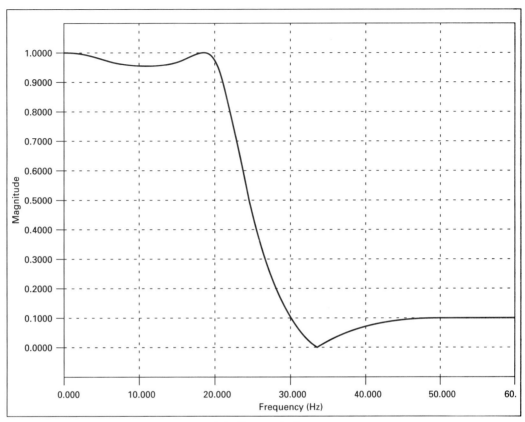

Figure B.9 Elliptic filter magnitude response.

B.4 ELLIPTIC FILTERS

An elliptic low-pass filter is characterized by the squared-magnitude function

$$|G(j\Omega)|^2 = \frac{1}{1 + \epsilon^2 \psi_N^2(\Omega/\Omega_c)}, \qquad (B.58)$$

where ϵ is a positive constant. The function $\psi_N(x)$ is called a *Chebyshev rational function* and is defined in terms of Jacobi elliptic functions. A thorough treatment of elliptic functions would be well beyond the scope of this text and will not be given. Detailed discussions of elliptic filter theory, related formulas, and their derivations can be found in references 1 and 15.

Typical elliptic filter magnitude behavior is illustrated in Fig. B.9. Elliptic filters exhibit ripple in both the passband and stopband. In return, for a given filter order, they provide sharper transition characteristics compared to Butterworth and Chebyshev filters.

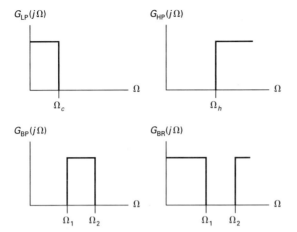

Figure B.10 Notational conventions used for frequency transformations.

B.5 FREQUENCY TRANSFORMATIONS

An analog low-pass prototype filter can be transformed into a high-pass, bandpass, or band-reject filter by means of a transformation applied to the *s*-domain transfer function. Low-pass to high-pass, low-pass to bandpass, and low-pass to band-reject filter transformation methods will be discussed in this section. Figure B.10 illustrates the notational conventions that will be used for representing the critical frequencies of the four filter types.

Low pass to high pass. A low-pass filter can be converted into a high-pass filter by means of the transformation

$$G_{HP}(s) = G_{LP}\left(\frac{\Omega_c \Omega_h}{s}\right). \quad (B.59)$$

The frequency Ω_c of the low-pass filter is mapped to the frequency Ω_h for the high-pass filter.

Low pass to bandpass. A low-pass filter can be converted into a bandpass filter by means of the transformation

$$G_{BP}(s) = G_{LP}\left(\frac{s^2 + \Omega_0^2}{s}\right), \quad (B.60)$$

where Ω_0 is the *geometric center frequency* of the passband of the resulting bandpass filter; that is,

$$\Omega_0 = \sqrt{\Omega_1 \Omega_2}. \quad (B.61)$$

Also, it can be shown that

$$\Omega_c = \Omega_2 - \Omega_1. \quad (B.62)$$

Sec. B.5 Frequency Tranformations

Low pass to band-reject. A low-pass filter can be converted into a band-reject filter by means of the transformation

$$G_{BR}(s) = G_{LP}\left(\frac{\Omega_0^2 s}{s^2 + \Omega_0^2}\right), \quad (B.63)$$

where Ω_0 is the *geometric center frequency* of the passband of the resulting bandpass filter; that is,

$$\Omega_0 = \sqrt{\Omega_1 \Omega_2}. \quad (B.64)$$

Also, it can be shown that

$$\frac{\Omega_1 \Omega_2}{\Omega_c} = \Omega_2 - \Omega_1. \quad (B.65)$$

References

1. A. Antoniou, *Digital Filters: Analysis and Design*, McGraw-Hill, New York, 1979.
2. R. E. Blahut, *Fast Algorithms for Digital Signal Processing*, Addison-Wesley, Reading, MA, 1985.
3. N. K. Bose, *Digital Filters Theory and Application*, Elsevier Science Publishing Co., New York, 1985.
4. C. S. Burrus and T. W. Parks, *DFT/FFT and Convolution Algorithms Theory and Implementation*, Wiley, New York, 1985.
5. B. A. Carlson, *Communication Systems*, McGraw-Hill, New York, 1985.
6. R. Kuc, *Introduction to Digital Signal Processing*, McGraw-Hill, New York, 1988.
7. E. P. Cunningham, *Digital Filtering: An Introduction*, Houghton-Mifflin, Boston, 1992.
8. A. G. Dezcky, "Synthesis of Recursive Digital Filters Using the Minimum P-Error Criterion," *IEEE Trans. on Audio ElectroAcoustics*, AU-20, no 4, pp. 257–263, 1972.
9. R. W. Hamming, *Digital Filters*, Prentice-Hall, Englewood Cliffs, NJ, 1988.
10. F. J. Harris, "On the Use of Windows for Harmonic Analysis with the Discrete Fourier Transform," *Proc. IEEE*, vol. 66, pp. 51–83, January 1978.
11. S. S. Haykin, *Adaptive Filter Theory*, Prentice-Hall, Englewood Cliffs, NJ, 1986.
12. IEEE Digital Signal Processing Committee, *Selected Papers in Digital Signal Processing II*, IEEE Press, New York, 1976.
13. IEEE Digital Signal Processing Committee, *Selected Computer Programs for Digital Signal Processing*, IEEE Press, New York, 1979.
14. L. B. Jackson, *Digital Filters and Signal Processing*, Kluwer Academic Publishers, Boston, 1986.
15. M. T. Jong, *Methods of Discrete Signal and System Analysis*, McGraw-Hill, New York, 1982.
16. J. F. Kaiser, "Nonrecursive Digital Filter Design Using the I_0-Sinh Window Function," *Proc. 1974 IEEE Int Symp. on Circuits and Systems*, pp. 20–23, April 1974.
17. J. H. McClellan and T. W. Parks, "A Unified Approach to the Design of Optimum FIR Linear-phase Digital Filters," *IEEE Trans on Circuit Theory* 20, no. 6, pp. 697–701, November 1973.
18. A. V. Oppenheim and R. W. Schafer, *Discrete-time Signal Processing*, Prentice-Hall, Englewood Cliffs, NJ, 1989.
19. T. W. Parks and C. S. Burrus, *Digital Filter Design*, Wiley, New York, 1987.

20. T. W. Parks and J. H. McClellan, "Chebyshev Approximation for Nonrecursive Digital Filters with Linear Phase," *IEEE Trans. on Circuit Theory* 19, no. 4, pp. 189–194, March 1972.
21. J. G. Proakis and D. G. Manolakis, *Introduction to Digital Signal Processing*, Macmillan, New York, 1988.
22. L. R. Rabiner and B. Gold, *Theory and Application of Digital Signal Processing*, Prentice-Hall, Englewood Cliffs, NJ, 1975.
23. R. A. Roberts and C. T. Mullis, *Digital Signal Processing*, Addison-Wesley, Reading, MA, 1987.
24. K. Steiglitz, "Computer-aided Design of Recursive Digital Filters," *IEEE Trans. on Audio ElectroAcoustics* AU-18, pp. 123–129, 1970.
25. R. D. Strum, and D. E. Kirk, *First Principles of Discrete Systems and Digital Signal Processing*, Addison-Wesley, Reading, MA, 1988.

Index

Add constant offset, 166
Add two sequences, 165
Addition of sequences, 20
Aliasing, 29, 75, 86, 90, 96, 97, 119, 127
All-pass system, 107, 108
Amplitude:
 characteristics, 72
 modulation, 23
Analog:
 filter design, 181
 prototype, 114, 118, 119, 120, 123, 128
 signals, 82
 system, 10
Analog-to-digital conversion, 82
Analyze filter, 183
Angular frequency, 14, 29, 102, 115
Anti-aliasing filter, 97
Aperiodic, 16
Approximation error, 50, 52
Arithmetic operations submenu, 27, 165, 170
ARMA, 188
Attenuation characteristics, 69
Autocorrelation, 175
Autocovariance, 175
AUTOEXEC.BAT, 153
Autoregressive, 186
Average power, 19

Band-limited interpolation, 88
Band limiting, 97
Bartlett window (*see* Triangular window)
Bessel function, 139, 162
Biased formula, 175, 176
BIBO stability criterion, 53
Bilinear transformation, 119, 124, 125, 126, 130, 182, 183
Blackman window, 138, 140, 162, 177, 178
Blackman-Tukey method, 186
Block diagram, 31, 112
Box-Muller, 163
Burg method, 187
Butterworth, 129, 181, 182, 190, 191

Carrier, 23
Cartesian form, 169
Cascade form, 112
Causal system, 108
Causality, 53, 55, 107, 113

Chebyshev, 178, 179, 181, 182, 190, 196
 inverse, 202
 polynomials, 196, 197, 198
Circular:
 convolution, 75, 78
 shift, 77, 78, 155, 167
Close window, 158
Closed forms, 17
Compander, 94, 171
Complex:
 conjugate, 167
 exponential sequence, 14
 plane, 101, 102
Compressor, 93, 94
CONFIG.SYS, 2, 152
Configuration:
 file, 157
 options submenu, 156, 158, 180
Conjugate:
 antisymmetric, 27, 67
 symmetric sequence, 27, 67
Convolution, 2, 36, 105, 137, 174
 circular, 75, 78
 graphical interpretation of, 37
 linear, 79
 superposition interpretation of, 40
 theorem, 64
Convolve sequences, 38
Copy, 22
 segment from sequence, 164
 sequence, 22, 163
Correlation, 2
Cost function, 114
Create new file, 158
Critical frequencies, 128, 130, 182
Cross correlation, 175
Cross covariance, 176
Cube-root, 52
Current:
 sequence, 158
 transform, 180
Cut segment from sequence, 164
Cutoff frequency, 121, 133, 136

Data directory, 156
Data entry form, 8
 formats, 188
Decimation in time, 178
Decimator, 95
Decomposition of a sequence, 26
Delete:
 filter, 181
 sequence, 159
 transform, 180

Deterministic signals, 19
DFT, 72, 73, 80, 81, 178
 inverse, 79, 179
Difference equation, 7, 46, 48, 106, 107, 113, 122, 125, 173
Differential equation, 122
Differentiation in frequency, 66, 105
Differentiator, 133, 150, 185
Digital signal, 82
Digital-to-analog conversion, 82
Direct-form, 112
Directories, 156
Discrete Fourier transform, 72
 inverse, 79, 179
Discrete-impulse sequence, 35
Discrete-time Fourier transform, 3, 6, 44, 58, 101
 properties of, 60
 symmetry properties of, 67
Discrete-time signals, 3, 31
 classification methods for, 16
 symmetry properties of, 26
Discrete-time sinusoids, 28
Discrete-time structures, 101, 111
Discrete-time systems, 11, 31
Discrete-to-continuous converter, 84, 87
Discrete-window functions, 162
Display mode, 157
Divide sequences, 166
Division by zero, 166
Don't care band, 115
Double-click delay, 157
Downsample sequence, 177
Downsampler, 95
Downsampling, 96, 97
Driving noise variance, 187
DTFT (*see* Discrete-time Fourier transform)
DTLTI system, 38, 46, 47, 105, 107, 111

Editor submenu, 157, 159
Elementary discrete-time signals, 13
Elementary operations, 13, 20
Elliptic, 124, 181, 182, 190, 205
Energy and power signals, 19
Enter external filter, 184
Equiripple, 149
Euler's formula, 18, 45, 60, 63
Even component, 26, 167
Even sequence, 26
Excess tolerance, 182, 183

211

Expander, 94
Exponential, 24, 163, 171

FFT, 177
Filter:
    analysis, 2
    design, 2
    implementation, 2
    order, 181, 182
Filters menu, 181
Finite precision, 13
Finite-impulse-response filters, 132
Finite-length sequence, 104
FIR:
    filter design submenu, 184
    filters, 132
First backward difference, 52, 119, 122, 123, 124, 127, 182, 183
First-order hold, 88, 89, 99
Flip sequence, 22, 167
Formula entry method, 4, 54, 76, 135, 160
Fourier series method, 133, 141, 184
Fourier transforms, 2
Frequency:
    shifting, 63
    synthesis, 106
    transformations, 129, 182, 190, 206
Frequency-domain, 57
    analysis of systems, 69
Frequency sampling design, 144, 147, 148, 185

Gain characteristics, 69
Gaussian, 20, 163
Generate sequence, 4
    submenu, 160
Geometric center frequency, 206, 207
Geometric series, 17
Gibbs phenomenon, 136, 137
Graphics, 2
    display, 1
GRAPHICS.COM, 2, 152
Group-delay, 69, 113, 183

Hamming window, 138, 140, 141, 143, 162, 177, 178
Hanning window, 138, 140, 162, 177, 178
Hardware requirements, 151
Hilbert transform filter, 133, 150, 185
Histogram, 177
Hyperbolic functions, 173

Ideal sampler, 83
IIR (*see also* Infinite-impulse-response):

filter design, 182
filters, 113
Imaginary part, 167
Import:
    ASCII data file, 159, 160
    MATLAB data file, 159
Impulse:
    invariance, 119, 121, 127, 130, 182, 183
    response, 46, 58, 106, 109, 113, 120, 183
    sequence, 163
Infinite-impulse-response, 3
    filters, 113
Initial conditions, 47
Insert segment into sequence, 164, 165
Interpolation, 145
    filter, 98
Interpolator, 95, 98
Inverse:
    DFT, 79, 179
    DTFT, 58
    FFT, 178
    z-transform, 108, 109, 180

Kaiser window, 138, 139, 140, 162, 177, 178
    design of FIR filters, 184

Laplace transform, 101, 120
Left-sided sequence, 104, 111
Left-sided term, 109
Limit sequence, 172
Linear:
    convolution, 79
    phase, 70, 132, 136, 138, 146
    shift, 155
    simultaneous equations, 144
    systems, 32, 57
Linearity, 32, 61, 76, 105, 106
List:
    filters, 181
    sequences, 158
    transforms, 180
Log files directory, 156
Logarithm, 24, 171
Long division, 109, 110, 111, 180, 181

Macro, 33
Macro files directory, 157
Magnitude, 169, 183
    characteristics, 72
    response, 69
Main lobe, 137, 139, 162
Make complex, 28, 164
Make periodic, 164
Matched z-transform, 119, 127, 130, 182, 183
Math coprocessor, 2, 152

Mean, 163, 175
Mean-square value, 92, 177
Median, 177
Midriser quantizer, 91
Midtread quantizer, 91
Modulation:
    index, 23
    theorem, 63
Modulo-N, 28, 81, 167, 168
Mouse options, 157
Moving average, 187
    filter, 4, 7, 53
Multiband filter, 148, 149, 150, 185
Multiplication:
    by a constant, 21
    of sequences, 20
Multiplicative congruential method, 163
Multiply:
    by constant, 166
    two sequences, 165

Newton's method, 51, 52
Next window, 158
Nonlinear operations submenu, 25, 169
Nonuniform quantization, 94
Normalized frequency, 14, 29, 115, 116
Nyquist rate, 86

Odd component, 26, 168
Odd sequence, 26
On-line help, 4, 151
Open existing file, 158
Operating system, 1
Operations menu, 165
Overlays, 2, 152

Parallel combination, 41
Parallel form, 112
Parks McClellan, 133, 149, 150, 185
Parseval's theorem, 66
Partial fraction expansion, 109, 120
Passband, 115
    tolerance, 115, 116, 128, 181, 182
PATH statement, 153
PC-DSP, 1, 2
Periodicity, 16
Periodogram, 186
Phase, 169, 183
    distortion, 70
    response, 69
Plot, 5
    sequence, 159
    transform, 180
Polar form, 169
Postprocessing, 55

Index

Power series, 180
Power spectrum estimation, 2
Previous window, 158
Prewarping, 128, 130
Probability density function, 92, 163
Processing functions submenu, 173
Progress monitor, 2, 4, 153, 164
Pseudo code, 113
Pull-down menu, 153

Quantization, 3, 83, 90, 91
 error, 92
 olevels, 172
 step size, 172
Quantize sequence, 6, 171, 172
Quit PC-DSP, 156

Radix-2, 178
Raise to power, 170
Raised cosine, 148, 162, 185
RAM (*see* Random-access memory)
RAM disk, 157
Ramp:
 function, 25
 sequence, 163
Random-access memory, 152
Random process, 175, 176
 sequences, 20, 27, 28, 162
 signals, 19
Read from keyboard, 71, 160
Real-part, 146, 167
Real-time, 49, 55
Reciprocate, 170
Reconstruction, 82, 84, 87
 filter, 88
Rectangular approximation, 50
Rectangular window, 73, 137, 138, 162, 177, 178
 spectrum of, 137
Reflection coefficient, 187
Region of convergence, 102, 103, 108
Repeat delay, 157
Right-sided exponential sequence, 14, 163
Right-sided sequence, 103, 104, 110, 111
Right-sided term, 109
Round-off errors, 24
Run macro, 86
Running integral, 49
 rectangular approximation of, 50
 trapezoidal approximation of, 50, 51
Running sum, 49, 50, 64

s-plane, 123
Sample, 11
 mean, 177
 variance, 177
Sampling, 82, 84
 frequency, 28, 84, 116
 interval, 28, 84
 period, 84
Sampling rate, 84, 95, 98, 116
 changing, 95
Save as:
 ASCII data file, 159
 MATLAB data file, 160
Save options, 157
Scalar addition, 21
Sequence, 4, 11, 154
 editing submenu, 163
Sequences menu, 158
Series combination, 42
Shift sequence, 22, 146, 154, 155, 166
Shuffling, 163
Side lobe, 137, 139, 162
Signal, 11
Signal statistics, 19, 177
Signal-system interaction, 35, 65
Signal-to-noise ratio, 93
Simulate filter, 184
Simulation, 87
Size/move window, 158
SNR (*see* Signal-to-noise ratio)
Specification diagram, 116, 117
Spectrum menu, 186
Square, 5, 170
Square-root, 51, 170
Stability, 53, 54, 107, 108, 113
Stable system, 109
Starting index, 160
Stationary, 175, 176
Statistics submenu, 177
Status line, 153, 154
Steady-state, 45
Step sequence, 163
Stopband, 115
 attenuation, 116, 136
 tolerance, 115, 128, 181, 182
Subtract sequences, 146, 165
Superposition, 39, 65
System:
 function, 43, 44, 46, 57, 101, 107
 menu, 146

Tabulate, 5
 sequence, 159
 transform, 180

Temporary directory, 157
Temporary exit to DOS, 156
Text editor, 157
Time:
 invariance, 34
 reversal, 21, 62, 105
 shifting, 21, 62, 105, 106
Time-delay, 69, 113, 183
Time-invariant systems, 32, 57
Transfer function, 106, 113
Transform analog filter, 183
Transformation, 31
Transforms menu, 6, 177
Transition band, 115, 146, 147, 185
Trapezoidal approximation, 50, 51, 125, 130
Triangular window, 138, 139, 142, 162, 177, 178
Trigonometric functions, 172
Two-sided sequence, 104

Unbiased formula, 175, 176
Uniform, 20, 163
Unit circle, 102
Unit-impulse sequence, 13
Unit-ramp sequence, 21
Unit-sample sequence, 13
Unit-step sequence, 14
Upsample sequence, 177
Upsampler, 95, 97
Upsampling, 99
 rate, 98
User-interface, 151

Variance, 163

Waveform:
 generator, 16, 25
 manipulation, 2
 synthesis, 2
Weighting function, 114
Welch's method, 186
Window functions, 73, 133

Yule-Walker method, 187

z-plane, 123
z-transform, 101
 inverse, 108, 109, 180
Zero padding, 74
Zero-crossing, 99
Zero-order hold, 88, 99
Zoom window, 158